New Frontiers in Cryptography

Khaled Salah Mohamed

New Frontiers in Cryptography

Quantum, Blockchain, Lightweight, Chaotic and DNA

 Springer

Khaled Salah Mohamed
A Siemens Business
Fremont, CA, USA

ISBN 978-3-030-58998-1   ISBN 978-3-030-58996-7   (eBook)
https://doi.org/10.1007/978-3-030-58996-7

© The Editor(s) (if applicable) and The Author(s), under exclusive license to Springer Nature
Switzerland AG 2020
This work is subject to copyright. All rights are reserved by the Publisher, whether the whole or part of
the material is concerned, specifically the rights of translation, reprinting, reuse of illustrations, recitation,
broadcasting, reproduction on microfilms or in any other physical way, and transmission or information
storage and retrieval, electronic adaptation, computer software, or by similar or dissimilar methodology
now known or hereafter developed.
The use of general descriptive names, registered names, trademarks, service marks, etc. in this publication
does not imply, even in the absence of a specific statement, that such names are exempt from the relevant
protective laws and regulations and therefore free for general use.
The publisher, the authors, and the editors are safe to assume that the advice and information in this book
are believed to be true and accurate at the date of publication. Neither the publisher nor the authors or the
editors give a warranty, expressed or implied, with respect to the material contained herein or for any
errors or omissions that may have been made. The publisher remains neutral with regard to jurisdictional
claims in published maps and institutional affiliations.

This Springer imprint is published by the registered company Springer Nature Switzerland AG
The registered company address is: Gewerbestrasse 11, 6330 Cham, Switzerland

Dedicated to the memory of my grandfather and grandmother.

Preface

Today, cryptography plays a vital role in every electronic and communication system. Everyday, many users generate and interchange large amount of information in various fields through internet, telephone conversations and e-commerce transactions. In modern system-on-chips (SoCs), cryptography plays an integral role in protecting the confidentiality and integrity of information. This book provides comprehensive coverage of various cryptography topics, while highlighting the most recent trends such as quantum, blockchain, lightweight, chaotic and DNA cryptography. Moreover, this book covers cryptography primitives and their usage and applications and focuses on the fundamental principles of modern cryptography such as stream ciphers, block ciphers, public key algorithms and digital signatures. The readers will build a solid foundation in cryptography and security. This book presents the fundamental mathematical concepts of cryptography. Moreover, this book presents hiding data techniques such as steganography and watermarking. Besides, it provides a comparative study of the different cryptographic methods that can be used efficiently to solve security problems. This book discusses modern cryptography and data hiding techniques. This includes:

- *Stream ciphers, block ciphers, public key algorithms and digital signatures.*
- *Quantum cryptography*: Quantum cryptography uses physics to develop a cryptosystem completely secure against being compromised without the knowledge of the sender or the receiver of the messages.
- *Blockchain cryptography*: Blockchain is a distributed database that allows direct transactions between two parties without the need for an authoritative mediator. Blockchain cryptography is a way to encapsulate transactions in the form of blocks where blocks are linked through the cryptographic hash, hence forming a chain of secured blocks.
- *Lightweight cryptography*: Lightweight cryptography works between the trade-offs of security, cost, and performance and is focused at devices and systems on edge.
- *Chaotic cryptography*: Many strong ciphers have been applied widely, such as DES, AES and RSA. But most of them cannot be directly used to encrypt

real-time embedded systems because their encryption speed is not fast enough and they are computationally intensive. So, chaotic cryptography is suitable for real-time embedded systems in terms of performance, area and power efficiency.

- *DNA cryptography*: DNA cryptography is a promising and rapid emerging field in data security. DNA cryptography may bring forward a new hope for unbreakable algorithms. DNA cryptology combines cryptology and modern biotechnology.
- *Steganography*: Steganography is the art of hiding the existence of a message between sender and intended recipient. It hides secret messages in various types of files such as text, images, audio and video.
- *Watermarking*: Watermarking is embedding some information within a digital media so that the digital media looks unchanged. This is useful for copy protection.

Fremont, CA, USA   Khaled Salah Mohamed

Contents

Chapter 1
Introduction to Cyber Security

Today, cryptography plays a vital role in every electronic and communication system. Everyday many users generate and interchange large amount of information in various fields through the Internet, telephone conversations, and e-commerce transactions. In modern system-on-chips (SoCs), cybersecurity plays an integral role in protecting the confidentiality and integrity of information. *Cybersecurity* is protecting computers, servers, mobiles, networks, electronic devices, and data from malicious attacks [1]. Recent years have seen an unfortunate and disruptive growth in the number of cyber-attacks. There are mainly three threats to data security [2, 3]:

- Theft them (confidentiality/privacy).
- Modify them (Integrity).
- You are prevented to get them (access/availability).

The aim of any secure system is to ban these threats. There are many techniques for achieving this such as encryption and data hiding. We will cover them in this book.

1.1 Security Terms

- *Confidentiality* refers to the protection of information, such as computer files or database elements, so that only authorized persons may access it in a controlled way [4].
- *Integrity* refers to not being able to modify information unless proper authorization is used. *Availability* refers to the presence of information when it is needed by authorized personnel and accessed using proper security measures.
- *Vulnerability* means weakness in the secure system.
- *Threat* is set of circumstances that have the potential to cause loss or harm.
- *Attack* is the act of a human exploiting the vulnerability in the system [5].

K. S. Mohamed, *New Frontiers in Cryptography*,
https://doi.org/10.1007/978-3-030-58996-7_1

- *Trojan horse* is software that appears to perform legitimately but has malicious side effect.
- *Virus* is a self-propagating Trojan horse; infects other software.
- *Worm* is a Virus which propagates over network.

1.2 Security Threats/Attacks

Security means freedom from risk or danger. Generally, nothing is ever 100% secured. Given enough time, resources, and motivation, an attacker can break any system. There are many threats to data security (Fig. 1.1):

- *Interception*: Theft them (confidentiality/*privacy attack*), i.e., *eavesdrop (nondestructive)*.
- *Modification*: Modify them (Integrity attack), i.e., *insert* messages into connection (destructive). *Hijacking by* taking over ongoing connection by removing sender or receiver, inserting himself as an attacker in place. It is also called fabrication of data or counterfeit data [6, 7].
- *Interruption*: You are prevented to get them (access/*availability attack*), i.e., *denial of service (DoS)* as attacker can prevent service from being used by others (e.g., by overloading resources).

Cybercrimes are criminal offenses committed via the Internet or otherwise aided by various forms of computer technology. There are many privacy concerns surrounding cybercrime when confidential information is intercepted or disclosed, lawfully or otherwise [8].

1.3 Security Requirements/Services/Objectives/Goals

Below we describe the main security requirements to overcome the security threats (Fig. 1.2):

Fig. 1.1 Security attacks and threats

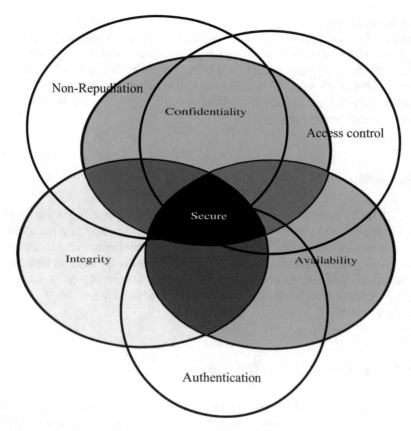

Fig. 1.2 Security goals intersections

1.3.1   Confidentiality

Refers to the protection of information, such as computer files or database elements, so that only authorized persons may access it in a controlled way. Confidentiality ensures that the message is encoded in order to conceal it, so the sender encrypts the message (plaintext) to create a ciphertext that is transmitted. The receiver, who possesses the cryptographic key, decrypts the ciphertext into the original plaintext.

1.3.2   Authentication

Authentication answers the following question "how does a receiver know that remote communicating entity is who it is claimed to be?". It is also called *identification*. Nowadays, most cryptographic algorithms support authenticated encryption

(AE) or authenticated encryption with associated data (AEAD). This basically means that both confidentiality and authenticity of the data is achieved. When referring to the AEAD scheme, it is assumed that the recipient is able to verify the integrity of both the encrypted and the decrypted message. To clarify this even more, the associated data (AD) are used to bind a ciphertext to the context that it is supposed to be. So, any attempt to place a valid ciphertext along with a different context is detectable and can be rejected.

1.3.3   Integrity

Refers to not being able to modify information unless proper authorization is used. The information and data sent can't be modified in storage or during the transmission between the source and destination in a way that the alteration is not detectable. Data integrity assures that the message received is exactly the same as the one sent by the sender. This may be accomplished, e.g., with the use of hash functions like SHA256 that create a unique digest from the original message, which is sent along with the message.

1.3.4   Access Control/Authorization

Who is allowed to do what. Access control is the process of controlling who does what and ranges from managing physical access to equipment to dictating who has access to a resource, such as a file, and what they can do with it, such as read or change the file. Many security vulnerabilities are created by the improper use of access controls.

1.3.5   Availability

Refers to the presence of information when it is needed by authorized personnel and accessed using proper security measures.

1.3.6   Non-repudiation

The ability to ensure that a party to a contract or a communication cannot deny the authenticity of their signature on a document or the sending of a message that they originated. You can't deny doing something you did. Generally, it is the assurance that the sender can't repudiate the validity of the message transmitted. This is accomplished with the use of digital signatures (especially used in online transactions) and message authentication codes, which are basically hash functions containing a key. It should be noted that such cryptographic primitives also ensure the integrity of the information, in a more robust manner than a simple hash function.

System is secured when all these goals are achieved (Fig. 1.2).

1.4   Security Mechanisms/Tools/Defenses

Security tools are summarized as below (Fig. 1.3):

- *Cryptographic algorithms* (Table 1.1): can be symmetric (one shared key) or asymmetric algorithms (we have two keys: one is secret, other is public) [9, 10].
- *Authentication:* who the user actually is. It is achieved by digital signature.
- *Public/private keys:* give out public key. Encrypt with this. Decrypt with private key.
- *Hashes:* create a unique, fixed length signature (hash) of a data set.
- *Digital signatures:* encrypt hash with private key. Decrypt with public key. Encryption does *not* ensure integrity.
- *Passwords:* something you know. It should be hard enough.
- *Firewalls:* a firewall is like a castle with a drawbridge. Only one point of access into the network.
- *Trusted third party:* a trusted third party can issue declarations such as "the holder of this key is a person who is legally known."

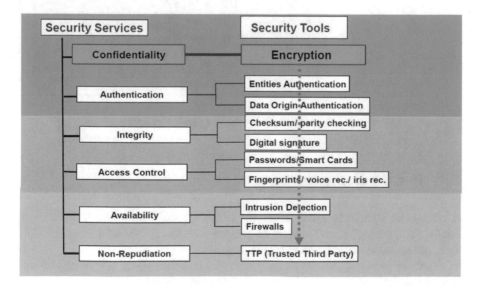

Fig. 1.3 Security services and tools

Table 1.1 Classifications of cryptographic algorithms

	Symmetric algorithms	Asymmetric algorithms (Public/private keys)	Hash algorithms
Examples	DES, AES, 3DES, RC5	RSA, ECC, DH, ECDH	MD4, HMAC, SHA-1
Math	Table lookup Permutations Multiplication Modular addition Modular multiplication Fixed shift/rotate Variable shift/rotate	Modular exponentiation Point multiplication on elliptic curves	Multiplication Addition Logical operations Fixed shift/rotate

Fig. 1.4 An example for security hierarchy

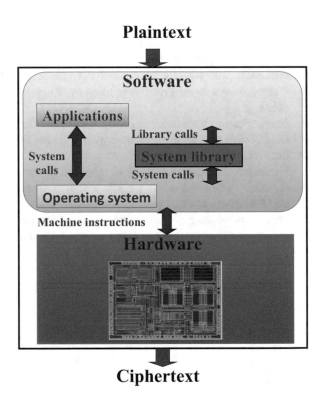

1.5   Security Hierarchy/Levels

A computing system is a collection of hardware (HW), software (SW), storage media, data, networks, and human interacting with them. We need to secure SW, data, and communication, and HW (Fig. 1.4).

Another prospective for the security hierarchy is shown in Fig. 1.5. Vulnerabilities can happen on the level of hardware, software, and data [11, 12].

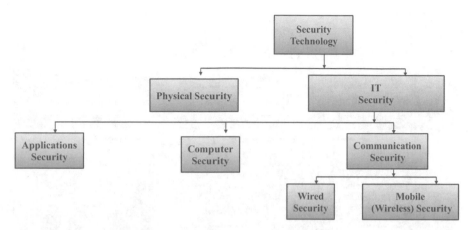

Fig. 1.5 Security hierarchy. Another prospective

- *Hardware Vulnerabilities:* Adding devices, changing them, removing them, intercepting the traffic to them, or flooding them with traffic until they can no longer function. Hardware vulnerabilities are often introduced by hardware design flaws. RAM memory, for example, is essentially capacitors installed very close to one another. It was discovered that, due to proximity, constant changes applied to one of these capacitors could influence neighbor capacitors. Based on that design flaw, an exploit called Rowhammer was created. By repeatedly rewriting memory in the same addresses, the Rowhammer exploit allows data to be retrieved from nearby address memory cells, even if the cells are protected.
- *Software Vulnerabilities:* Software can be replaced, changed, or destroyed maliciously, or it can be modified, deleted, or misplaced accidentally. Whether intentional or not, these attacks exploit the software's vulnerabilities. *Malware* is any code that can be used to steal data, bypass access controls, or cause harm to, or compromise a system such as spyware and Ransomware. Software vulnerabilities are usually introduced by errors in the operating system or application code; despite all the effort companies put into finding and patching software vulnerabilities, it is common for new vulnerabilities to surface. Microsoft, Apple, and other operating system producers release patches and updates almost every day. Application updates are also common. Applications such as web browsers, mobile apps, and web servers are often updated by the companies or organizations responsible for them. *Phishing* is when a malicious party sends a fraudulent email disguised as being from a legitimate, trusted source. The message intent is to trick the recipient into installing malware on their device or into sharing personal or financial information. An example of phishing is an email forged to look like it was sent by a retail store asking the user to click a link to claim a prize. The link may go to a fake site asking for personal information, or it may install a virus.
- *Data Vulnerabilities:* Confidential data leaked to a competitor.

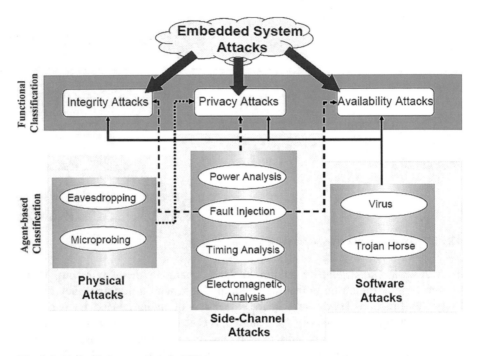

Fig. 1.6 Embedded system attacks [13]

Example for different types of embedded system attacks are shown in Fig. 1.6. IP protection is also an important topic. Without IP protection, companies can lose revenue and market share.

IP vendors are facing major challenges to protect hardware IPs from IP piracy as, unfortunately, recent trends in IP piracy and reverse engineering efforts to produce counterfeit ICs have raised serious concerns in the IC design community. IP piracy can take several forms, as illustrated by the following:

1. A chip design house buys an IP core from an IP vendor and makes an illegal copy or "clone" of the IP. The IC design house then sells it to another chip design house (after minor modifications) claiming the IP to be its own.
2. An untrusted fabrication house makes an illegal copy of the GDS-II database supplied by a chip design house and then illegally sells them as hard IP.
3. An untrusted foundry manufactures and sells counterfeit copies of the IC under a different brand name.
4. An adversary performs post silicon reverse engineering on an IC to manufacture its illegal clone.

These scenarios demonstrate that all parties involved in the IC design flow are vulnerable to different forms of IP infringement which can result in loss of revenue and market share. Hence, there is a critical need of a piracy proof design flow that equally benefits the IP vendor, the chip designer, as well as the system designer. A desirable characteristic of such a secure design flow is that it should be transparent to the end user, i.e., it should not impose any constraint on the end user with regard to its usage, cost, or performance.

To secure an IP, we need to obfuscate it then encrypt the contents before sending it to the customer. *Obfuscation* is a technique that transforms an application or a design into one that is functionally equivalent to the original but is significantly more difficult to reverse engineer. So, obfuscation changes the name of all signals to numbers and characters combination. The second level is to encrypt the whole files. Although encryption is effective, code obfuscation is an effective enhancement that further deters code understanding for attackers. Moreover, *watermarking* can be used to protect soft-IPs. It includes modules duplication or module splitting. For ASIC, circuit camouflage is used. *Circuit camouflage* lets individual logic cells appear identical at each mask layer, when in fact subtle changes are present to differentiate logic functions. Changes are designed so that the reverse engineer is unable to automate cell recognition. To protect PCB, we encapsulate the PCB into epoxy (black blobs) and add a few fake layers for complexity.

1.6   Mathematical Background

1.6.1   Modular Arithmetic

In modular arithmetic, we map the product of any computation (addition, multiplication) to a bounded set of integers. The bound is defined by the *modulus* (or base). Let a, r, m integers and $m > 0$, then

$a \equiv r \bmod m$, if m divides $a - r$. An example is $43 \equiv 1 \bmod 7$.

1.6.2   Greatest Common Divisor

Assume we need to find gcd (r_0, r_1). A solution is to factor r_0, r_1. Then, the gcd should be the highest common factor. Factoring is complicated and hard for large numbers. An example is shown below:

$r_0 = 72 = 2 \times 2 \times 3 \times 7$ and $r_1 = 24 = 2 \times 3 \times 4$ then gcd $(r_0, r_1) = 6$.

1.7   Security Protocols

A protocol is a series of steps carried out by two or more entities. A security protocol is a protocol that runs in an untrusted environment and tries to achieve a security goal. Examples for security protocol are *IPSec* and *SSL*. Nowadays, for data encryption through the Internet, the HTTPS (Hypertext Transfer Protocol

Secure) protocol uses SSL/TLS-based encryption to create a secure channel to shared data. The cryptographic protocols TLS (Transport Layer Security) and SSL (Secure Sockets Layer) used by HTTPS use asymmetric cryptography which uses a pair of keys for sending information, authenticating the receiver more reliably.

1.7.1   SSL

It is widely deployed security protocol and supported by almost all browsers, web servers. Moreover, it is available to all TCP applications [14]. SSL operates at the *presentation layer* in the OSI model (Layer6) as depicted in Fig. 1.7 [15]. Public key encryption lies at the heart of Secure Sockets Layer, which is another common form of encryption on the Internet. In an SSL connection, your computer and the target computer take the roles of the two correspondents, swapping public keys and encoding all data that travels back and forth between the two machines. This ensures that file transfers and other communications remain secure, although outsiders may still be able to determine the nature of the transfer by looking at public packet information – a packet's destination port, for example, may give away the type of transfer, as most Internet protocols use easily identifiable port numbers.

1.7.2   IPSec

It is an Internet Engineering Task Force (IETF) standard suite of protocols between two communication points across the IP network that provide data authentication, integrity, and confidentiality. It also defines the encrypted, decrypted, and authenticated packets [16, 17]. SSL and IPSec both boast strong security pedigrees with comparable throughput speed, security, and ease of use for most customers of commercial VPN services [18]. IPSec works on *network layer* from the OSI layers (Fig. 1.7).There's a hierarchy of seven levels in the OSI model, namely:

1. The physical layer
2. The data link layer
3. The network layer
4. The transport layer
5. The session layer
6. The presentation layer
7. The application layer

IPSec encryption can create significant network bottlenecks, whereas Layer 2 encryption introduces virtually no latency or overhead to the network.

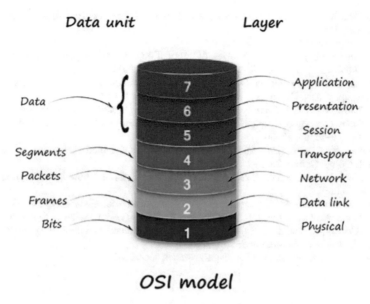

Fig. 1.7 Open System Interconnection (OSI) model

1.8 Conclusions

This chapter discusses the fundamentals of cryptography. It provides a comprehensive study of the three critical aspects of security: confidentiality, integrity, and authentication. Cryptography plays a vital and critical role in achieving the primary aims of security goals, such as authentication, integrity, confidentiality, and non-repudiation. Cryptographic algorithms are developed in order to achieve these goals.

References

1. C. Paar, J. Pelzl, *Understanding Cryptography: A Textbook for Students and Practitioners* (Springer, London, UK, 2009)
2. C. Paar, J. Pelzl, B. Preneel, *Understanding Cryptography: A Textbook for Students and Practitioners* (Springer Heidelberg Dordrecht, Bochum, 2010)
3. R.A. Mollin, *Codes: The Guide to Secrecy from Ancient to Modern Times* (Chapman and Hall/CRC, Boca Raton, 2005)
4. S. Pincock, *Codebreaker: The History of Codes and Ciphers*, 1st edn. (Walker, 2016). ISBN: 978-0802715470
5. William Stallings, *Cryptography and Network Security Principles and Practices* (Pearson Education Inc, 2006)
6. Ç.K. Koç, *Cryptographic Engineering* (Springer, 2009)

7. K.M. Martin, *Everyday Cryptography Fundamental Principles and Applications* (Oxford University Press, 2012)
8. https://www.avast.com/c-cybercrime
9. N. Sharma, Prabhjot, H. Kaur, A review of information security using cryptography technique. Int. J. Adv. Res. Comp. Sci. **8**(Special Issue), 323–326 (2017)
10. S. Tayal, N. Gupta, P. Gupta, D. Goyal, M. Goyal, A review paper on network security and cryptography. Adv. Comp. Sci. Tech. **10**(5), 763–770 (2017)
11. P.P. Charles, P.L. Shari, *Security in Computing*, 4th edn. (Prentice-Hall, Inc, 2008)
12. S. William, *Cryptography and Network Security: Principles and Practice*, 2nd edn. (Prentice-Hall, Inc, 1999), pp. 23–50
13. U. Guin, N. Asadizanjani, M. Tehranipoor, Standards for hardware security. GetMobile: Mobile Comp. Comm. **23**, 5–9 (2019)
14. https://www.csoonline.com/article/3246212/what-is-ssl-tls-and-how-this-encryption-protocol-works.html
15. https://security.stackexchange.com/questions/19681/where-does-ssl-encryption-take-place
16. S. Bellovin, *Problem Areas for the IP Security Protocols* (Usenix Security Symposium, 1996)
17. C. McCubbin, A. Selcuk, D. Sidhu, Initialization vector attacks on the IPsec protocol suite, in *IEEE Workshop on Enterprise Security*, (2000)
18. https://www.comparitech.com/blog/vpn-privacy/ipsec-vs-ssl-vpn/

Chapter 2
Cryptography Concepts: Confidentiality

2.1 Cryptography History

The term cryptography is derived from the Greek word Kryptos. Kryptos is used to describe anything that is hidden, veiled, secret, or mysterious. Cryptography is the study of mathematical techniques for the secure transmission of a private message over an insecure channel. Cryptography is the basic technique to secure our data from different kind of attackers like: interruption, modification, fabrication, etc. It has been around for 2000+ years. They were shaving the slave's head, tattooed the message on it, and let the hair grow [1]. The first documented use of cryptography in writing dates back to circa 1900 B.C. when an Egyptian scribe used nonstandard hieroglyphs in an inscription. Some experts argue that cryptography appeared spontaneously sometime after writing was invented, with applications ranging from diplomatic missives to war-time battle plans. Caesar's cipher: shifting each letter of the alphabet by a fixed amount. It is easy to break. Vigenere's polyalphabetic cipher generalizes Caesar's shift cipher. It uses keyword to select encrypting rows. *Substitution cipher*: permutations of 26 letters, using the dictionary. It is easy to break. Then in modern era, we have many cryptography algorithms as we will explain in the rest of this chapter. Modern cryptography algorithms are based over the fundamental process of factoring large integers into their primes, which is said to be intractable [2, 3]. In both *symmetric* and *asymmetric* cryptosystems, encryption is the process of changing the original form of text to unreadable form and decryption process gets the original form of data from the meaningless text. For block cipher, use of plaintext and ciphertext of equal size avoids data expansion. In this chapter, confidentiality will be addressed. Encryption is the major tool to achieve confidentiality [4–6].

K. S. Mohamed, *New Frontiers in Cryptography*,
https://doi.org/10.1007/978-3-030-58996-7_2

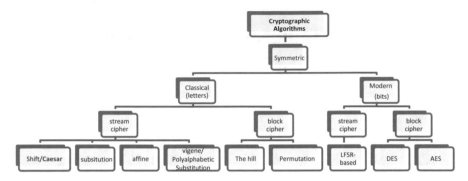

Fig. 2.1 Symmetric cryptographic algorithm. Symmetric means sender and receiver have a shared secret key

2.2 Symmetric Encryption

Figure 2.1 shows a summary for different symmetric encryption algorithms. Symmetric means sender and receiver have a shared secret key. There are two primitive operations with which strong encryption algorithms can be built: *confusion* and *diffusion*. In confusion, an encryption operation where the relationship between key and ciphertext is obscured, and a common element for achieving confusion is substitution. In diffusion, an encryption operation where the influence of one plaintext symbol is spread over many ciphertext symbols with the goal of hiding statistical properties of the plaintext and a common element for achieving diffusion is through permutations (i.e., transposition). *Avalanche effect* is considered as one of the desirable property of any encryption algorithm. A slight change in either the key or the plaintext should result in a significant change in the ciphertext.

2.2.1 Historical Algorithms: Letter-Based Algorithms

2.2.1.1 Caesar Cipher

This is one of the oldest and earliest examples of cryptography, invented by Julius Caesar, the emperor of Rome, during the Gallic Wars. In this type of algorithm, the letters A through W are encrypted by being represented with the letters that come three places ahead of each letter in the alphabet, while the remaining letters A, B, and C are represented by X, Y, and Z. This means that a "shift" of 3 is used, although by using any of the numbers between 1 and 25, we could obtain a similar effect on the encrypted text. Therefore, nowadays, a shift is often regarded as a Caesar Cipher (Fig. 2.2) [7]. A Python code for Caesar cipher is shown in Fig. 2.3.

Fig. 2.2 Caesar cipher
Encryption: C = P + K
mod 26   (1)
Decryption: P = C − K
mod 26   (2)

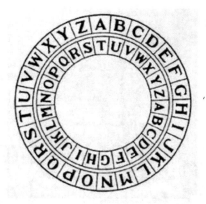

```
1    # Caesar Cipher
2
3    # Encryption process
4    input_plaintext_file  = open("plaintext", "rt")
5    encrypted_word_file   = open("encrypted", "wt")
6    encrypted_word        = []
7    final_encrypted_words = []
8    for line in input_plaintext_file:
9        for eachletter in line:
10           if eachletter == " " :
11               encrypted_word.append ( eachletter.replace(eachletter, eachletter))
12           else:
13               encrypted_word.append ( eachletter.replace(eachletter, chr (ord(eachletter) +1 )))
14               final_encrypted_words = ''.join(encrypted_word)
15           encrypted_word_file.write( str(final_encrypted_words) )
16   input_plaintext_file.close()
17   encrypted_word_file.close()
18
19
20   # decryption process
21   input_encrypted_file  = open("encrypted", "rt")
22   output_plaintext_file = open("decrypted", "wt")
23   decrypted_word        = []
24   final_decrypted_words = []
25   for line in input_encrypted_file:
26       for eachletter in line:
27           if eachletter == " " :
28               decrypted_word.append ( eachletter.replace(eachletter, eachletter))
29           else:
30               decrypted_word.append ( eachletter.replace(eachletter, chr (ord(eachletter) -1 )))
31               final_decrypted_words = ''.join(decrypted_word)
32           output_plaintext_file.write( str(final_decrypted_words) )
33   input_encrypted_file.close()
34   output_plaintext_file.close()
```

Fig. 2.3 Python code for Caesar cipher example: shift by 1 letter

2.2.1.2 Substitution Ciphers

In a substitution cipher, we take the alphabet letters and place them in random order under the alphabet written correctly. In the encryption and decryption, the same key is used. The rule of encryption here is that "each letter gets replaced by the letter beneath it," and the rule of decryption would be the opposite [8]. The substitution cipher did not shift the data by a fixed number. Instead, it shifts it by a random one. A Python example for substitution cipher is shown in Fig. 2.4.

```
1   # Subsitution Cipher
2
3   #Key Generation
4   import random
5   from random import seed
6   seed (1)
7   # Encryption process
8   input_plaintext_file   = open("plaintext", "rt")
9   encrypted_word_file    = open("encrypted", "wt")
10  encrypted_word         = []
11  final_encrypted_words = []
12  for line in input_plaintext_file:
13      for eachletter in line:
14          if eachletter == " " :
15              encrypted_word.append ( eachletter.replace(eachletter, eachletter))
16          else:
17              Key= random.randint(1, 26)
18              encrypted_word.append ( eachletter.replace(eachletter, chr ((ord(eachletter) + Key )% 127 )))
19              final_encrypted_words = ''.join(encrypted_word)
20      encrypted_word_file.write( str(final_encrypted_words) )
21  input_plaintext_file.close()
22  encrypted_word_file.close()
23
24
25  # decryption process
26  seed (1)
27  input_encrypted_file   = open("encrypted", "rt")
28  output_plaintext_file = open("decrypted", "wt")
29  decrypted_word         = []
30  final_decrypted_words = []
31  for line in input_encrypted_file:
32      for eachletter in line:
33          if eachletter == " " :
34              decrypted_word.append ( eachletter.replace(eachletter, eachletter))
35          else:
36              Key= random.randint(1, 26)
37              decrypted_word.append ( eachletter.replace(eachletter, chr ((ord(eachletter) - Key ) % 127) ))
38              final_decrypted_words = ''.join(decrypted_word)
39      output_plaintext_file.write( str(final_decrypted_words) )
40  input_encrypted_file.close()
41  output_plaintext_file.close()
```

Fig. 2.4 Python code for substitution cipher example

2.2.1.3 Transposition/Polyalphabetic Ciphers

Instead of substituting letters, rearrange them. Transposition can be defined as the alteration of the letters in the plaintext through rules and a specific key. A columnar transposition cipher can be considered as one of the simplest types of transposition cipher and has two forms: the first is called "complete columnar transposition," while the second is "incomplete columnar." Regardless of which form is used, a rectangle shape is utilized to represent the written plaintext horizontally, and its width should correspond to the length of the key being used. There can be as many rows as necessary to write the message. When complete columnar transposition is used, the plaintext is written, and all empty columns are filled with null so that each column has the same length [9]. The key cannot have a repetitive letters. Python code for transposition cipher example is shown in Fig. 2.5. The algorithm can be summarized as follows:

1. The message is written out in rows of a fixed length and then read out again column by column, and the columns are chosen in some scrambled order according to the key.
2. Width of the rows and the permutation of the columns are usually defined by a keyword.

```
1   # Transposition Cipher
2   # Key Generation
3   import math
4   key = "khaled"
5   # Encryption Function
6   def encrypt(msg):
7       encrypted_txt = ""
8       k_indx = 0
9       msg_len = float(len(msg))
10      msg_lst = list(msg)
11      key_lst = sorted(list(key))
12      col = len(key)
13      row = int(math.ceil(msg_len / col))
14      fill_null = int((row * col) - msg_len)
15      msg_lst.extend(' ' * fill_null)
16      matrix = [msg_lst[i: i + col]    for i in range(0, len(msg_lst), col)]
17      for i in range(col):
18          curr_idx = key.index(key_lst[k_indx])
19          encrypted_txt += ''.join([col[curr_idx]   for col in matrix])
20          k_indx += 1
21      return encrypted_txt
22  # Decryption Function
23  def decrypt(encrypted_txt):
24      msg = ""
25      k_indx = 0
26      msg_indx = 0
27      msg_len = float(len(encrypted_txt))
28      msg_lst = list(encrypted_txt)
29      col = len(key)
30      row = int(math.ceil(msg_len / col))
31      key_lst = sorted(list(key))
32      dec_cipher = []
33      for i in range(row):
34          dec_cipher += [[None] * col]
35      for i in range(col):
36          curr_idx = key.index(key_lst[k_indx])
37          for j in range(row):
38              dec_cipher[j][curr_idx] = msg_lst[msg_indx]
39              msg_indx += 1
40          k_indx += 1
41      msg = ''.join(sum(dec_cipher, []))
42      return msg
43  # Run Code
44  encrypted_word_file     = open("encrypted", "wt")
45  output_plaintext_file   = open("decrypted", "wt")
46  with open("plaintext") as infile:
47      for line in infile:
48          msg = ''.join (list(line))
49  encrypted_txt = encrypt(msg)
50  encrypted_word_file.write( format(encrypted_txt) )
51  plain_after_decrypt =  decrypt(encrypted_txt)
52  output_plaintext_file.write( format(plain_after_decrypt) )
53  encrypted_word_file.close()
54  output_plaintext_file.close()
```

Fig. 2.5 Python code for transposition cipher example

3. For example, the key "KHALED" is of length 6, and the permutation is defined by the alphabetical order of the letters in the keyword. In this case, the order would be "3 6 5 2 1 4."
4. Any spare spaces are filled with nulls or left blank.

5. Finally, the message is read off in columns, in the order specified by the keyword.
6. To decipher it, the recipient has to work out the column lengths by dividing the message length by the key length.
7. Then, write the message out in columns again, then re-order the columns by reforming the key word.

2.2.1.4 One-Time Pad Cipher

The cryptographic algorithms that already exist have the common strategy to have a large key space and a complicated algorithm. For symmetric cryptography, the use of one time pad is the simplest solution to the key distribution problem. However, with increasing advancement in technology, it is getting easier to break the algorithms that are widely in use. The increasing length of OTPs is also a cause of concern. Polyalphabetic cipher continued for almost 400 years. In 1882, Frank Miller invented cipher called as one-time pad. In one-time pad, a key was selected whose length was same as that of the plaintext message. The shifts in the plaintext never followed a repetitive pattern, and the encrypted message had uniform frequency distribution, thereby providing no leakage of the information. One-time pad (OTP) cipher is unbreakable due to the following [10]:

• The key is as long as the given message.
• The key is truly random and especially auto-generated.
• Key and plain text calculated as modulo 10/26/2.
• Each key should *be used once* and destroyed by both sender and receiver.
• There should be two copies of key: one with the sender and other with the receiver.

The problem with the one-time pad is that, in order to create such a cipher, its key should be as long as or even longer than the plaintext. In other words, if you have 500 megabyte video file that you would like to encrypt, you would need a key that's at least 4 gigabits long.

2.2.2 Modern Algorithms: Bits-Based Algorithms

2.2.2.1 Stream Ciphers

Stream ciphers operate on pseudorandom bits generated from the key, and the plaintext is encrypted by XORing both the plaintext and the pseudorandom bits. Stream ciphers use conceptual tools similar to block ciphers. Substitution is the primary tool: each bit or byte of plaintext is combined with the key material by an Exclusive-OR (XOR) operation to substitute the plaintext bit into the ciphertext bit.

Binary XOR is quite simple. There are only two possible values, 1 and 0, and if the two inputs are the same, the result is 0; otherwise it is 1 [11]. Stream cipher's security depends entirely on the "suitable" key stream, while randomness plays a main role, so the random number generator (RNG) is significant for that purpose. *Neural cryptography* is a new source for public key cryptography schemes which are not based on number theory and have less computation time and memory complexities. Neural cryptography can be used to generate a common secret key between two parties [12].

E0 Cipher

E0 is a stream cipher used in the Bluetooth protocol. It generates a sequence of pseudorandom numbers and combines it with the data using the XOR operator. The key length may vary but is generally 128 bits. At each iteration, E0 generates a bit using four shift registers of differing lengths (25, 31, 33, 39 bits) and two internal states, each 2 bits long. At each clock tick, the registers are shifted, and the two states are updated with the current state, the previous state and the values in the shift registers. Four bits are then extracted from the shift registers and added together. The algorithm XORs that sum with the value in the 2 bit register. The first bit of the result is output for the encoding [13].

2.2.2.2   Block Ciphers

In a block cipher, two values are generally referred to: the size of the block and the size of the key. The security relies on the value of both. Many block ciphers use a 64 bit block or a 128 bit block. As it is crucial that the blocks are not too large, the memory footprint and the ciphertext length are small in size. Regarding the ciphertext length, blocks instead of bits are processed in a block cipher. Block ciphers can be symmetric or nonsymmetric as will be discussed later in this chapter.

We start with *plaintext*. Something you can read. We apply a mathematical algorithm to the plaintext.

The algorithm is the *cipher*. The plaintext is turned in to *ciphertext*. In symmetric encryption, same key is used to encrypt and decrypt the data. The shared key K between sender and receiver should be kept secret. The encryption/decryption process can be modeled by eq. (2.1) and can be seen in Fig. 2.6:

$$P = D_K \left(E_k \left(P \right) \right) \tag{2.1}$$

Symmetric encryption is fast to encrypt and decrypt, suitable for large volumes of data. Comparative analysis of symmetric encryption algorithm is shown in Table 2.1.

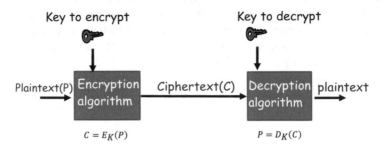

Fig. 2.6 Symmetric key encryption

Table 2.1 Comparative analysis of symmetric encryption algorithm [14, 15]

Algorithms/ parameters	DES	3DES	AES	Blowfish
Published	1977	1998	2001	1993
Developed by	IBM	IBM	Vincent Rijmen, Joan Daeman	Bruce Schneier
Algorithm structure	Feistel	Feistel	Substitution-Permutation	Feistel
Block cipher	Binary	Binary	Binary	Binary
Key length	56 bits	112 bits, 168 bits	128 bits, 192 bits and 256	32–448 bits
Flexibility or modification	No	Yes, extended from 56 to 168 bits	Yes, 256 key size is multiple of 64	Yes, 64–448 key size in multiple of 32
Number of rounds	16	48	10, 12, 14	16
Block size	64 bits	64 bits	128 bits	64 bits
Throughput	Lower than AES	Lower than DES	Lower than Blowfish	High
Level of security	Adequate security	Adequate security	Excellent security	Excellent security
Encryption speed	slow	Very slow	Fast	Fast
Effectiveness	Slow in both software and hardware	Slow in software	Effective in both software and hardware	Efficient in software
Attacks	Brute force attack	Brute force attack, known plaintext, chosen plaintext	Side channel attack	Dictionary attack

Rijndael Algorithm

Rijndael is a symmetric key encryption algorithm that's constructed as a block cipher. It supports key sizes of 128, 192, and 256 bits, with data handling taking place in 128 bit blocks. In addition, the block sizes can mirror those of their respective keys. This last specification puts Rijndael over the limits required for AES design conditions, and the Advanced Encryption Standard itself is looked upon as a subset of the Rijndael algorithm [16].

SAFER Algorithm

In cryptography, SAFER (Secure And Fast Encryption Routine) is the name of a family of block ciphers designed primarily by James Massey (one of the designers of IDEA) on behalf of Cylink Corporation. The first SAFER cipher was SAFER K-64, published by Massey in 1993, with a 64 bit block size. The "K-64" denotes a key size of 64 bits. There was some demand for a version with a larger 128 bit key. This algorithm is of interest for several reasons. It is designed for use in software. Unlike DES, or even IDEA, it does not divide the block into parts of which some parts affect others; instead, the plaintext is directly changed by going through S-boxes, which are replaced by their inverses for decryption. SAFER uses eight rounds. The first step for a round is to apply the first subkey for the round to the 8 bytes of the block. The operation by which each byte of the subkey is applied to each byte of the block depends on which byte is used: the sequence is XOR, add, add, XOR, XOR, add, add, XOR. Then, the S-box is used. Those bytes to which the subkey was applied by an XOR go through the regular S-box; those bytes to which it was applied by addition go through the inverse S-box [17].

AES Algorithm

Data Encryption Standard (DES) was the encryption standard till 2001 when it was replaced by AES. The Advanced Encryption Standard (AES) was established by National Institute of Standards and Technology (NIST) in 2001 as the current standard for encrypting electronic data [18]. AES is based on *Rijndael* cipher which is an iterated block cipher with a fixed block length and supports variable key lengths. A block length of 128 bits and three different key sizes of 128, 192, and 256, which require 10, 12, and 14 rounds, respectively, are used. Figures 2.7 and 2.8, respectively, show AES ciphering and deciphering. AES is working as follows (Fig. 2.9):

1. Add round key: round key is XORed with the plaintext, and then the result will be converted into 4× 4 *matrix* (state).
2. The sub-byte transformation is a nonlinear *substitution* operation that works on bytes. It's based on the Galois field GF (2^8) with irreducible polynomial $m(x) = x^8 + x^4 + x^3 + x + 1$ and can be done using look up table.

Fig. 2.7 AES ciphering

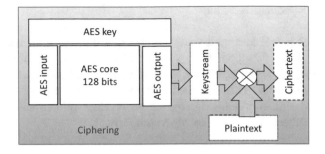

Fig. 2.8 AES deciphering

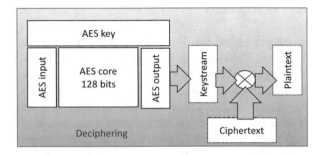

3. The shift row transformation *rotates* each row of the input state to the left; the first row will remain unshifted; the second row will be rotated by 1 step to the left, whereas the second row will be rotated by two steps; and the third row will be rotated 3 steps to the left.
4. Mix columns: each input column is considered as a polynomial over GF (2^8) and *multiplied* with the constant polynomial $a(x) = \{03\}x3 + \{01\}x^2 + \{01\}x + \{02\}$ modulo $x^4 + 1$. The coefficients of $a(x)$ are also elements of GF (2^8) and are represented by hexadecimal values in this equation. The inverse mix column transformation is the multiplication of each column with $a(x) = \{0B\}x3 + \{0D\}x^2 + \{09\}x + \{0E\}$ modulo $x^4 - 1$.
5. Key expansion.

Blowfish

Blowfish is a 64 bit symmetric key block cipher used to efficiently encrypt data. The Blowfish algorithm has a Feistel structure which means that it iterates a simple encryption function 16 times. It uses variable key lengths ranging from 32 bits up to a maximum of 448 bits. It was first introduced by the security guru Bruce Schneier in 1993, and it has never been cracked [19]. Twofish, also designed by Schneier, is the successor of Blowfish, and it contains a 128 bit block along with key up to 256 bits long [20].

Fig. 2.9 AES encryption

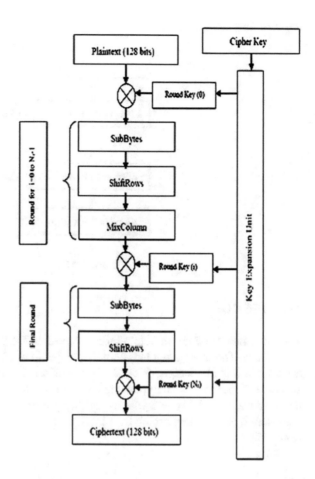

Des

The Data Encryption Standard (DES) is a symmetric key block cipher published by the National Institute of Standards and Technology (NIST). DES is an implementation of a Feistel cipher. It uses 16 round Feistel structure. The block size is 64 bit. Though, key length is 64 bit, DES has an effective key length of 56 bits, since 8 of the 64 bits of the key are not used by the encryption algorithm (function as check bits only). DES structure is shown in Fig. 2.10 [21]. *Feistel cipher* is not a specific scheme of block cipher. It is a design model from which many different block ciphers are derived. DES is just one example of a Feistel cipher. A cryptographic system based on Feistel cipher structure uses the same algorithm for both encryption and decryption. The encryption process uses the Feistel structure consisting multiple rounds of processing of the plaintext, each round consisting of a "substitution" step followed by a permutation step.

Fig. 2.10 DES structure

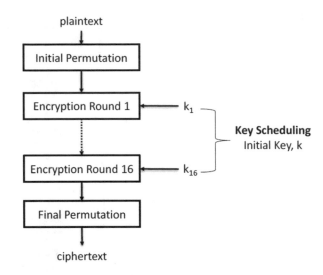

Triple DES/3DES

The basic methodology of Triple DES is based on DES. It encrypts the plain text three times. The overall key length of Triple DES is 192 bits, i.e., three 64 bit keys are used in this block cipher. Like DES, the effective length of each key is 56 bits. In Triple DES, the first encrypted cipher text is again encrypted by the second key, and again the third key encrypts the resulting encrypted cipher text. Though this algorithm is much more secure than DES, it is too slow for many real life applications [22].

RC4 (Rivest Cipher 4)

This is a stream cipher published in 1987. The disadvantage of these techniques is the generation of the key stream, which is a potentially long sequence of values consisting of 40 bit to 128 bit key and a 24 bit initialization vector. But the encryption is straightforward; the actual plaintext is xor'd with the key stream. The key input is a pseudorandom bit generator that creates a byte number, which is the output of the generator and it is called the key stream. The key stream is then XORed with the plaintext byte by byte to produce the ciphertext. It is impossible to predict the key stream without having knowledge of the key.

RC5 (Rivest Cipher 4)

This is a block cipher algorithm, published in 1994, which uses variable block size, key size, and encryption steps. The block size can be 32, 64, or 128 bits, and the key size can be from 0 up to 2040 bits. Additionally, the number of rounds can be from 0 up to 255. It consists of some modular additions and exclusive OR (XOR) operations. It is vulnerable to a differential attack using numerous chosen plaintexts, where the inputs differences can also affect the outputs [23].

SEAL Algorithm

SEAL is a length-raising "illusive random" that depicts 32 bit string N- to L-bit string SEAL under a hidden 160 bit key. The output length L is intended to be diverse, however, in general bound to 64 kilobytes. The key usage is to figure out three secret charts: R, S, and T; these charts have 256, 256, and 512 32 bit values, respectively, that are induced from the Secure Hash Algorithm (SHA). SEAL is the fruit of the dual shower source clarified. The first generator implements a systematic relies on the deducted charts R and T. It maps the 32-bit string n and the 6-bit counter [24].

2.3   Asymmetric Encryption

Many applications use asymmetric cryptography to secure communications between two parties. One of the main issues with asymmetric cryptography is the need for vast amounts of computation and storage.

The concept of public key cryptography evolved from an -attempt to attack two of the most difficult problems associated with symmetric encryption. The first problem is that of key distribution, where under symmetric encryption requires either that two communicants already share a key, which somehow has been distributed to them or the use of a *key distribution center.* Asymmetric algorithms rely on one key for encryption and a different but related key for decryption [25]. There are many methods for key establishment, including certificates and public key infrastructure (*PKI*).

The *public key* is freely distributable. It is related mathematically to the private key, but you cannot (easily) reverse engineer the *private key* from the *public key*. Use the *public key* to encrypt data.

Only someone with the *private key* can decrypt.

The key for encryption (K_E) and decryption (K_D) are different. But, K_E and K_D form a unique pair. One of keys is made public and another made private. The encryption/decryption process can be modeled by eq. (2.2) and can be seen in Fig. 2.11.

$$P = D_{K_D}\left(E_{K_E}\left(P\right)\right) \tag{2.2}$$

Public key systems can be used for encryption and authentication. One key is used to encrypt the document; a different key is used to decrypt it. Comparison between symmetric and nonsymmetric encryption is shown in Table 2.2.

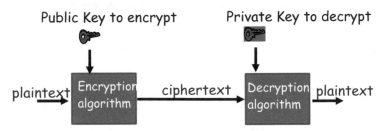

Fig. 2.11 Asymmetric key encryption

Table 2.2 Comparison between symmetric and nonsymmetric encryption

Symmetric encryption	Nonsymmetric encryption
1. The same algorithm with the same key is used for encryption and decryption	1. One algorithm is used for encryption and decryption with a pair of keys, one for encryption and one for decryption
2. The sender and receiver must share the algorithm and the key	2. The sender and receiver must each have one of the matched pair of keys (not the same one)
3. The key must be kept secret	3. One of the two keys must be kept secret
4. It must be impossible or at least impractical to decipher a message if no other information is available	4. It must be impossible or at least impractical to decipher a message if no other information is available
5. Knowledge of the algorithm plus samples of ciphertext must be insufficient to determine the key	5. Knowledge of the algorithm plus one of the keys plus samples of ciphertext must be insufficient to determine the other key

The security of asymmetric encryption rests on computational problems such as the difficulty of factorizing large prime numbers and the discrete logarithm problem. Such kind of algorithms is called one-way functions because they are easy to compute in one direction, but the inversion is difficult. Public key encryption works very well and is extremely secure, but it's based on complicated mathematics. Because of this, your computer has to work very hard to both encrypt and decrypt data using the system. In applications where you need to work with large quantities of encrypted data on a regular basis, the computational overhead means that public key systems can be very slow.

2.3.1 RSA: Factorization Computational Problem

The Rivest–Shamir–Adleman (RSA) scheme has been published in 1978, and since that time it is the most widely accepted and implemented general-purpose approach to public key encryption [26]. Encryption and decryption are of the following form, for some plaintext block M and ciphertext block C:

$$C = M^e \bmod n \tag{2.3}$$

$$M = C^d \bmod n = \left(M^e\right)^d \bmod n = M^{ed} \bmod n \tag{2.4}$$

Both sender and receiver must know the value of n. Plaintext is encrypted in blocks, with each block having a binary value less than some number n. That is, the block size must be less than or equal to $\log_2(n)$. The RSA scheme is a block cipher in which the plaintext and ciphertext are integers between 0 and n-1 for some n. The sender knows the value of e, and only the receiver knows the value of d. Thus, this is a public key encryption algorithm with a public key of = {e, n} and a private key of = {d, n}.

For this algorithm to be satisfactory for public key encryption, the following requirements must be met:

1. It is possible to find values of e, d, and n such that M^{ed} mod $n = M$ for all $M < n$. This is done by choosing $n = pq$, where p and q are primes numbers. Select e, where $1 < e < (p-1)\,(q-1)$ and $gcd\,((p-1)\,(q-1),\,e) = 1$, which means e is a prime to $(p-1)\,(q-1)$. Select d, where $d = e^{-1} mod\,((p-1)\,(q-1))$ or $de = 1\,mod((p-1)\,(q-1))$.
2. It is relatively easy to calculate mod M^e mod n and C^d for all values of $M < n$.
3. It is infeasible to determine d given e and n.

An example is shown below:

Assume plaintext = 88.
Key generation: $n = 17 \times 11 = 187$. $e = 7$, where gcd (160, 7) =1. $D = 23$, where $d*7 = 1$ mod 160.
Encryption process: $C = 88^7$ mod $187 = 11$.
Decryption process: $M = 11^{23}$ mod $187 = 88$.

2.3.2   ECC: Discrete Logarithm Problem (DLP)

Elliptic curve cryptography (ECC) is a public key cryptographic technique based on the algebraic structure of elliptic curves over finite fields. ECC is an approach to asymmetric cryptography used widely in low computation devices such as wireless sensor networks (WSNs) and Internet of Things (IoT) devices as it decreases power consumption and increases device performance [27]. This is due to its ability to generate small keys with a strong encryption mechanism. For example, encryption using the RSA algorithm with a 1024 bit key is equal to ECC encryption with a 160 bit key [28]. ECC ensures security depending on the ability to compute a point multiplication with a random point, as well as the inability to figure out a multiplicand given the original curve and product points. ECC uses a pair $(x; y)$ that fits into the equation $y^2 = x^3 + ax + b$ mod p together with an imaginary point (theta) at infinity, where a; $b \in Zp$ and $4a^3 + 27b^2 \neq 0$ mod p. ECC needs a cyclic Group G and the primitive elements we use, or pair elements, to be of order G.

2.3.3 ElGamal Cryptosystem

ElGamal cryptosystem is an asymmetric key encryption which was proposed by Taher Elgamal in 1984. The security of this algorithm is based on the discrete logarithm problem. For a given number, there is no existing algorithm which can find its discrete logarithm in polynomial time, but the inverse operation of the power can be derived efficiently. Another key aspect of ElGamal cryptosystem is randomized encryption. This algorithm can establish a secure channel for key sharing and generally used as key authentication protocol. For security the key size of this algorithm should be greater than 1024 bits. The major drawback of ElGamal algorithm is that it is relatively time-consuming.

2.3.4 Diffie–Hellman Algorithm: Key Exchange

Diffie–Hellman (DH) key exchange algorithm is a method for securely exchanging cryptographic keys over a public communications channel. Keys are not actually exchanged, but they are jointly derived. Traditionally, secure encrypted communication between two parties required that they first exchange keys by some secure physical means, such as paper key lists transported by a trusted courier. The Diffie–Hellman key exchange method allows two parties that have no prior knowledge of each other to jointly establish a shared secret key over an insecure channel. This key can then be used to encrypt subsequent communications using a symmetric key cipher. The steps are as follows [29]:

1. Alice and Bob publicly agree to use a modulus p and base g (which is a primitive root modulo p).
2. Alice chooses a secret integer a and then sends Bob $A = g^a$ mod p.
3. Bob chooses a secret integer b and then sends Alice $B = g^b$ mod p.
4. Alice computes $s = B^a$ mod p.
5. Bob computes $s = A^b$ mod p.
6. Alice and Bob now share a secret (s).

The difficulty of breaking these cryptosystems is based on the difficulty in determining the integer r such that $g^r = x$ mod p. The integer r is called the discrete logarithm problem of x to the base g, and we can write it as $r = \log_g^x$ mod p. The discrete logarithm problem is a very hard problem to compute if the parameters are large enough.

2.3.5   EGC

Elliptic curve cryptography (ECC) is an approach to public key cryptography based on the algebraic structure of elliptic curves over finite fields. ECC allows smaller keys compared to non-EC cryptography (based on plain Galois fields) to provide equivalent security. Elliptic curves are applicable for key agreement, digital signatures, pseudorandom generators and other tasks. Indirectly, they can be used for encryption by combining the key agreement with a symmetric encryption scheme. The EGC protocol generated high levels of data security to serve the purpose of protecting data during transmission in the IoT [30–32].

2.4   Hybrid Encryption

One of the known methods to strengthen the secureness of a cryptography is by combining two existing cryptographies. A combination of two cryptographies is also called the hybrid algorithm such as combination of symmetric and public key-based system. Symmetric key cryptography is faster and more efficient than public key cryptography but lacks security when exchanging keys over unsecured channels. Hybrid cryptosystems combine the speed of symmetric key cryptography with the security of public key cryptography. A hybrid cryptosystem consists of a public key cryptosystem for key encapsulation and a symmetric key cryptosystem for data encapsulation. Hybrid cryptosystems are used by most computer users in the form of HTTP Secure (*HTTPS*) [33].

2.5   Crypto-Analysis/Attacks

The analysis of the cryptography algorithms in general is known as cryptanalysis and is an essential aspect of testing the reliability of the cryptography system for practical. Cryptographic algorithms are provably secure against mathematical cryptanalysis under the black box assumption. Cryptanalysis methods are summarized below, and different types of attacks on different security levels are shown in Table 2.3. Attacks can be passive or active. *Passive attacks* do not alter or affect at any other way the information, and they do not cause any issue to the communication channel. The main goal here is to acquire unauthorized access to sensitive and confidential information and data. Passive attacks are often called as stealing information. What really makes this attack harmful is the fact that most of the times the owner is not aware that an unauthorized person has knowledge of the owner's data. For instance, an attacker could intercept and eavesdrop a communication channel and gain knowledge to confidential information, and neither the sender nor the receiver could figure that out. *Active attacks* is able to process the information and

Table 2.3 Different types of attacks on different security levels

Security abstraction level	Security objective	Side channel attack
Protocol	Authenticated communications	Man-in-the-middle, Traffic analysis
Algorithm	Encryption/hashing	Known plaintext, Known cryptext
Architecture	Functional integration (SW)	Stack smashing
Micro-architecture	Architecture integration (HW)	Bus probing
Circuit	Implementation	Differential power analysis

alter it in many different ways. More specifically, the attacker could change specific fields of the data like the originator name and the timestamp and generally modifying the information in an unauthorized way. Moreover, unauthorized deletion of data, initiation of unintended transmission of information or data, and, lastly, denial of access to data by legitimate users the so-called denial of service (DoS) attack are also examples of active attacks [34, 35].

2.5.1   Exhaustive/Brute Force Attack

The secret *key's space* should be long. Crypto-analysis time versus key size is shown in Table 2.4. The security of an encryption algorithm ought to have vast key space more sensitive to the secret key to tackling a different kind of attacks such as statistical attacks, differential attack, known plain text attacks, and exhaustive attacks. The large size of the key space also makes brute force attacks infeasible. Moreover, the *sensitivity of algorithms* toward the secret key during encryption and decryption is the key point of the robustness of an encryption algorithm. The higher the sensitivity, the more secure is the information because only a slight change in the key will lead toward an entirely different cipher image. That means no one can recover the original image except having the correct secret key [36–38].

2.5.2   Statistical/Histogram Attack

The histogram is a common approach to get the distribution of an image pixel values. Histogram of data should be uniform after encryption. This leads to statistical attacks invalid. In information theory, entropy is the most significant feature of the disorder. We can say numerical property reflecting the randomness associated unpredictability of an information source called entropy. The ideal *entropy* value for a random image with a gray level of 2^8 is 8. Which means the closer the entropy value is, the more is the haphazardness of an image, conclusively less information disclosed by the encryption scheme.

Table 2.4 Crypto-analysis time versus key size

Key size	Possible no. of keys	Time to crack	
		(1 encryption/μs)	(10^6 encryptions/ μs)
32	10^9	36 min	2. msec
56	10^{16}	1100 years	10 hrs
128	10^{38}	5×10^{24} years	5×10^{18} years
26 characters	10^{26}	6×10^{12} years	6×10^6 years

2.5.3   Differential Attack

Attackers often make a slight change to the original data and use the proposed algorithm to encrypt for the original data before and after changing, through comparing two encrypted image to find out the relationship between the original data and the encrypted data.

2.5.4   Known/Chosen Plaintext/Ciphertext Attack

2.5.4.1   Known Ciphertext Attacks (KCA)

Here the attacker possesses multiple ciphertexts, but without the corresponding plaintexts. This attack becomes effective when the corresponding plaintext can be extracted from one or more ciphertexts. Additionally, sometimes the encryption key can be discovered by this attack. In practice, however, the adversary performing this attack has also some knowledge about the plaintext. This information could be the language that the plaintext is written or the foreseeable statistical distribution of the characters in it.

2.5.4.2   Chosen Plaintext Attack (CPA)

The adversary here has free access to the encryption process and can create any ciphertext from any plaintext of his choice. So basically, the attacker can have any desirable pair of plaintext–ciphertext. This makes the process of finding the encryption key easier, as the attacker can gain more knowledge of the encryption operation, the more pairs of messages, and ciphertexts created.

2.5.4.3   Known Plaintext Attack (KPA)

This attack is quite similar to the previous one. Here the attacker knows the plaintext that the sender has sent and the corresponding ciphertext. The goal of the adversary is to gain information by taking advantage of the ciphertext–plaintext pairs they have. This could result to the discovery of the encryption key or other information for the algorithm as well. The difference with the chosen plaintext attack is that the plaintext is not chosen by the attacker but the sender of the message.

2.5.4.4   Chosen Ciphertext Attack (CCA)

In this type of attack, the adversary or the cryptanalyst has the ability to analyze any chosen ciphertexts along with the corresponding plaintexts. The goal is to gain the secret key or as much information as possible for the attacked cryptographic system. This attack holds with the assumption that the attacker can make the victim decrypt any encrypted message and send it to him. The more decrypted ciphertexts the attack owns, the more information is gained for the system, and thus it is more likely to break it.

2.5.4.5   Side Channel Attack

The system is attacked via the channel leaked information such as *time consumption, power consumption, or electromagnetic radiation*. Timing attack is one type of side channel attack. The attacker can access the equipment or physically damage them by performing, for instance, Differential Power Analysis (DPA) attack. In Differential Power Analysis, the attacker send lots of plaintext (bits) to the FPGA, which will decrypt them accordingly, and meanwhile the attacker will be measuring the power traces, trying to get the cryptography algorithm key (using statistical techniques and knowledge of the CMOS power model). There are many countermeasures against this attack, including changing the time of the key transmission or encryption to confuse the adversary and filtering the power line conditioning to prevent power-monitoring acts.

Deep learning technique can be used in side channel analysis context. Like other machine learning techniques, a deep learning technique builds a profiling model for each possible value of the targeted sensitive variable during the training phase, and, during the attack phase, these models are involved to output the most likely key used during the acquisition of the attack traces. In side channel attack context, an adversary is rather interested in the computation of the probability of each possible value deduced from a key hypothesis. Therefore, to recover the good key, the adversary computes the maximum or the log maximum likelihood approach like for template attack [39–41].

2.5.4.6   Man-in-the-Middle Attack

This attack depends on standing between the two communicating parties to get the message from the sender, change, and add to it and then send it forward to the receiver. This needs that the attacker knows the encryption keys to be able to encrypt the added parts to the message. One-time pad keys and changing the block lengths make the attacker not able to succeed to play the man-in-the middle role.

2.6 Comparison Between Different Encryption Algorithms

Stream ciphers are generally faster than block ciphers, due to the fact that in the second category, each block needs to be processed, one by one in order to be encrypted, which is not the case in stream ciphers where only one bit or byte is processed at a time. As a result, block ciphers require more memory allocation, since they have to work on bigger chunks of data, and, sometimes, they have to continue the operation from previous blocks as well. On the contrary, stream ciphers process at most a byte at a time, so they have relatively low memory requirements, and as a result they are cheaper to implement in constrained devices like embedded systems and IoT devices and more general in lightweight cryptographic algorithms. Nevertheless, stream ciphers are more difficult to develop and design effectively, and they are vulnerable depending on the usage. Stream ciphers do not offer integrity protection and authentication, whereas some block ciphers depending on the mode they use are able to provide integrity together with confidentiality. Additionally, because of the fact that block ciphers encrypt a whole block at a time and most of them have feedback modes, they are prone to adding noise in the transmission that could alter the data, so the rest of the transmission will not be the appropriate for the algorithm. Stream ciphers do not face such problem as the bits or bytes are encrypted separately from the other data, and most of the times there are solutions in case of connection issues. A comparative evaluation of different encryption algorithms is shown in Tables 2.5 and 2.6. DES algorithm has the lower encryption and decryption speed. AES algorithm has the least memory utilization. RSA has a very high memory utilization [42, 43]. Due to its low power consumption, DES is designed to work better in hardware than in software program. 3DES requires higher time than DES and consumes more power consumption and has less throughputs. The 3DES algorithm is the best for data protection because it uses three keys to encrypt and decrypt data [44].

Table 2.5 Comparison between different encryption algorithms

Algorithm	Encryption and decryption speed	Scalability	Vulnerabilities	Security
RSA	High	Not scalable	Brute force attack	High
DES	Low	Scalable	Brute force attack	Medium
AES	Low	Not scalable	Cryptanalysis attack	High
3DES	Low	Not scalable	Man-in-the-middle-attack	High
BLOWFISH	Low	Scalable	Brute force attack	Low
ECC	Low	Scalable	Brute force attack	Medium

Table 2.6 Comparison between AES, DES, and RSA

Factors	AES	DES	RSA
Development year	2000	1977	1978
Key length (bits)	128,192,256	56	1024
Key type	Symmetric	Symmetric	Asymmetric
Block size (bits)	128	64	512
Execution time	Fast	Moderate	Slow
Rounds	10,12,14	16	1

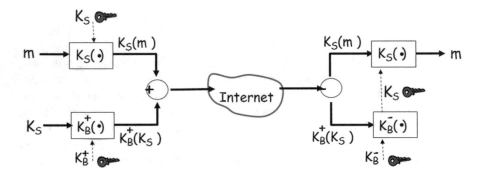

Fig. 2.12 Sending a secured email

2.7 Cryptography Applications

The need for strong cryptographic algorithms is very high, and the design of them is challenging, especially by taking into consideration that the processing power of computers increases day by day and malicious parties tend to find new vulnerabilities and breaches in security systems all the time. Nowadays, cryptography is excessively used in many types of applications.

2.7.1 Secured Email

Alice wants to send confidential email "m" to Bob (Fig. 2.12).
   Alice:

- Generates random symmetric private key, K_S
- Encrypts message with K_S
- Encrypts K_S with Bob's public key
- Sends both $K_S(m)$ and $K_B(K_S)$ to Bob

   Bob:

- Uses his private key to decrypt and recover K_S
- Uses K_S to decrypt $K_S(m)$ to recover m

2.7.2   Secured Chat

The chat client will use an encryption algorithm to provide end-to-end encrypted communications with other clients. There are many popular session setup protocol that powers real-world chat systems such as Signal and WhatsApp.

2.7.3   Secured Wireless Communication System

A high-level wireless communication system is shown in Fig. 2.13, where encryption plays a vital role. The *Bluetooth* encryption system uses the stream cipher E0 to encrypt the payloads of the packets which is re-synchronized for every payload. The E0 stream cipher consists of the payload key generator, the key stream generator, and the encryption/decryption part. The input bits are combined by the payload key generator and are shifted to the four linear feedback shift registers (LSFR) of the key stream generator. The key stream bits are then generated which are used for encryption. The Exclusive-OR operation is then performed on the key stream bits and data stream bits to generate the ciphertext. Similarly the Exclusive-OR operation is performed on the ciphertext to get back the plaintext during the decryption process.

WLAN encryption methods includes:

- Wired Equivalent Privacy (*WEP*): Wired Equivalent Privacy (WEP) is a security algorithm for IEEE 802.11 wireless networks. Standard 64 bit WEP uses a 40 bit key (also known as WEP-40), which is concatenated with a 24 bit initialization vector (IV) to form the RC4 key. At the time that the original WEP standard was drafted, the US Government's export restrictions on cryptographic technology limited the key size. Once the restrictions were lifted, manufacturers of access points implemented an extended 128 bit WEP protocol using a 104 bit key size (WEP-104).
- Wi-Fi Protected Access (*WPA*): Wi-Fi Protected Access (WPA) was the Wi-Fi Alliance's direct response and replacement to the increasingly apparent vulnerabilities of the WEP standard. WPA was formally adopted in 2003, a year before WEP was officially retired. The most common WPA configuration is WPA-PSK (pre-shared key). The keys used by WPA are 256 bit, a significant increase over the 64 bit, and 128 bit keys used in the WEP system [45].

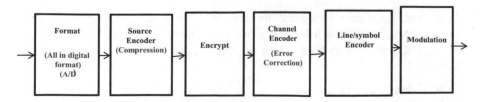

Fig. 2.13 High-Level wireless communication system

- Wi-Fi Protected Access 2 (*WPA2*): WPA has, as of 2006, been officially super-seded by WPA2. One of the most significant changes between WPA and WPA2 is the mandatory use of AES algorithms and the introduction of CCMP (Counter Mode Cipher Block Chaining Message Authentication Code Protocol) as a replacement for TKIP. However, TKIP is still preserved in WPA2 as a fallback system and for interoperability with WPA.

5G uses 256 bit encryption, a substantial improvement on the 128 bit standard used by 4G. With 5G, the user's identity and location are encrypted, making them impossible to identify or locate from the moment they get on the network [46]. In older 2G cellular systems, the cryptographic algorithms used to secure the air interface and perform subscriber authentication functions were not publicly disclosed. The GSM algorithm families pertinent to our discussion are A3, A5, and A8. A3 provides subscriber authentication, A5 provides air interface confidentiality, and A8 is related to A3, in that it provides subscriber authentication functions but within the SIM card. UMTS introduced the first publicly disclosed cryptographic algorithms used in commercial cellular systems. The terms UEA (UMTS Encryption Algorithm) and UIA (UMTS Integrity Algorithm) are used within UMTS as broad categories. UEA1 is a 128 bit block cipher called KASUMI, which is related to the Japanese cipher MISTY. UIA1 is a message authentication code (MAC), also based on KASUMI. UEA2 is a stream cipher related to SNOW 3G, and UIA2 computes a MAC based on the same algorithm. LTE builds upon the lessons learned from deploying the 2G and 3G cryptographic algorithms. LTE introduced a new set of cryptographic algorithms and a significantly different key structure than that of GSM and UMTS. There are three sets of cryptographic algorithms for both confidentiality and integrity termed EPS Encryption Algorithms (EEA) and EPS Integrity Algorithms (EIA). EEA1 and EIA1 are based on SNOW 3G, very similar to algorithms used in UMTS. EEA2 and EIA2 are based on the Advanced Encryption Standard (AES) with EEA2 defined by AES in CTR mode (e.g., stream cipher) and EIA2 defined by AES-CMAC (cipher-based MAC). EEA3 and EIA3 are both based on a Chinese cipher ZUC. While these new algorithms have been introduced in LTE, network implementations commonly include older algorithms for backward compatibility for legacy devices and cellular deployments [47].

2.7.4 Secured Mobile/Smartphone

Smartphone users are exposed to different threats. These threats may disturb operation and transfer user data from smartphones. There are a number of ways in which the type of attacks in mobiles can be segregated. Some of them are explained as follows:

- *Wi-Fi-based attacks*
  An attacker can intercept a Wi-Fi communication by doing eavesdropping. The security in WLAN is more vulnerable. It is possible for an attacker to break the

password easily get in the local network of the victim. In an event where an attacker prospers in breaking identification cipher, it becomes possible to attack both the phone and the entire network.

- *Web browser-based attack*
  In web browser-based attacks, an attacker uses leverages like stack-based overflow and other vulnerabilities in libraries. This is possible in all kinds of operating system either Android or iOS. Smartphones are also vulnerable to phishing and other malicious web site-based attacks, and the biggest problem with smartphones is that they don't have strong antivirus protection yet.
- *Operating system-based attacks*
  One may apply any number of secure mechanisms, but if there is vulnerability in operating system, it might be going to affect one day surely. There are several loopholes in operating systems of smartphones as these are in earlier stages and developers are not much aware about the kind of attacks possible. It was likely to dodge the security of operating system and circumvent the bytecode verifier and gain access of core operating system. Similarly in windows mobile OS, one can easily edit the general configuration file to a modifiable file. It is also possible for a malicious attacker to do modifications in the directory whenever an application is installed as at that time it has root privileges.

2.8 Conclusions

This chapter discusses the fundamentals of private and public key cryptography. Moreover, it explains the details of the main building blocks of these cryptographic systems. Besides, this chapter explores the different crypto-analysis techniques. It addresses stream ciphers, the Data Encryption Standard (DES) and 3DES, the Advanced Encryption Standard (AES), block ciphers, the RSA cryptosystem, and public key cryptosystems based on the discrete logarithm problem, elliptic curve cryptography (ECC), key exchange algorithms, and so many other algorithms. Moreover, this chapter provides a comparison between different encryption algorithms in terms of speed encryption, decoding, complexity, the length of the key, structure, and flexibility.

References

1. W. Stallings, L. Brown, *Computer Security: Principles and Practice* (Pearson, Harlow, 2019)
2. W. Stallings, *Cryptography and Network Security: Principles and Practice*, 5th edn. (Prentice Hall, Upper Saddle River, 2011). https://www.williamstallings.com
3. W. Stallings, L. Brown, *Computer Security: Principles and Practice*, 3rd edn. (Pearson Education, Boston, 2015). http://www.williamstallings.com/ComputerSecurity
4. P.V. Rao, H.M. Mallikarjun, Nagendra, S. Manjusha, Design and ASIC implementation of triple data encryption and decryption standard algorithm. Int. J. Power Elec. Tech. **1**(1), 1–15 (2011)

5. Y.-K. Lai, L.-G. Chen, J.-Y. Lai, Tai-Ming, VLSI architecture design and implementation for TWOFISH block cipher, in *Circuits and Systems, 2002*, (ISCAS, 2002)
6. X. Zheng, Z. Liu, B. Peng, Design and Implementation of an Ultra Low Power RSA Coprocessor, in *2008 4th International Conference on Wireless Communications, Networking and Mobile Computing, Dalian*, (2008), pp. 1–5
7. http://www.cs.trincoll.edu/~crypto/historical/caesar.html
8. https://www.sciencedirect.com/topics/computer-science/substitution-cipher
9. https://en.wikipedia.org/wiki/Transposition_cipher
10. https://www.tutorialspoint.com/cryptography_with_python/cryptography_with_python_one_time_pad_cipher.htm
11. https://www.sciencedirect.com/topics/computer-science/stream-ciphers
12. https://www.researchgate.net/publication/338556941_Single_secret_image_sharing_scheme_using_neural_cryptography
13. https://en.wikipedia.org/wiki/E0_(cipher)
14. M. Stamp, *Information Security: Principles and Practice* (Wiley, 2011)
15. P. Patil, P. Narayankar, D.G. Narayan, S.M. Meena, A comprehensive evaluation of cryptographic algorithms: DES, 3DES, AES, RSA and blowfish. Proc. Comp. Sci. **78**, 617–624 (2016)
16. https://blog.finjan.com/rijndael-encryption-algorithm/
17. http://www.quadibloc.com/crypto/co040301.htm
18. National Institute of Standards and Technology (NIST), Advanced Encryption Standard (AES). FIPS-197, 2001
19. A. Priya Sunny, B., and N, Design of blowfish encryption engine. Int. J. Comp. Sci. Tech. (IJCST) **5**(3), 46–48 (2014)
20. Wikipedia, Blowish (cipher) (2019), https://en.wikipedia.org/wiki/Blowfish_(cipher). Accessed 12 Dec 2019
21. K.D. Deepak, D. Pawan, Performance comparison of symmetric data encryption techniques. ISSN: 2278–1323. Int. J. Adv. Res. Comput. Eng. Tech. **1**(4) (2012)
22. J. Henry, 3DES is officially Being Retired, Cryptomathic (2018), https://www.cryptomathic.com/news-events/blog/3des-is-officially-being-retired. Accessed 12 Dec 2019
23. Wikipedia, RC5 (2019), https://en.wikipedia.org/wiki/RC5. Accessed 12 Dec 2019
24. B. Schneier, *Applied Cryptography Second Edition: Protocols, Algorithms, and Source* (China Machine Press, Beijing, 2000), pp. 239–252
25. D. Debasis, M. Rajiv, Programmable cellular automata based efficient parallel AES encryption algorithm. Int. J. Network Sec. Appl. (IJNSA) **3**(6), 204 (2011)
26. K. Parsi, S. Sudha, Data security in cloud computing using RSA algorithm. Int. J. Res. Comp. Comm. Tech. (IJRCCT), ISSN 2278- 5841, **1**(4), 145 (2012)
27. K.S. Mohamed, The era of internet of things: Towards a smart world, in *The Era of Internet of Things*, (Springer, Cham, 2019), pp. 1–19
28. P. Das, C. Giri, An efficient method for text encryption using elliptic curve cryptography, in *Proceeding of IEEE 8th International Advanced Computing Conference (IACC)*, (2018), pp. 96–101
29. https://mathworld.wolfram.com/Diffie-HellmanProtocol.html
30. D. Hankerson, A. Menezes, S.A. Vanstone, *Guide to Elliptic Curve Cryptography* (Springer-Verlag, 2004)
31. I. Blake, G. Seroussi, N. Smart (eds.), *Advances in Elliptic Curve Cryptography* (London Mathematical Society 317, Cambridge University Press, 2005)
32. L. Washington, *Elliptic Curves: Number Theory and Cryptography* (Chapman & Hall/CRC, 2003)
33. A.H. Vinck, *Introduction to Public Key Cryptography* (Duisburg-Essen, 2012). Retrieved from http://www.exp-math.uni-essen.de/~vinck/crypto/script-crypto-pdf/add-to-3.pdf
34. B. Sun, Z. Liu, V. Rijmen, R. Li, L. Cheng, Q. Wang, H. Alkhzaimi, C. Li, Links among impossible differential, integral and zero correlation linear cryptanalysis, in *Annual Cryptology Conference*, (Springer, 2015), pp. 95–115

35. S. Tajik, E. Dietz, S. Frohmann, J.-P. Seifert, D. Nedospasov, C. Helfmeier, C. Boit, H. Dittrich, Physical characterization of arbiter pufs, in *International Workshop on Cryptographic Hardware and Embedded Systems*, (Springer, 2014), pp. 493–509

36. C. Li, D. Lin, B. Feng, J. Lü, F. Hao, Cryptanalysis of a chaotic image encryption algorithm based on information entropy. IEEE Access **6**, 75834–75842 (2018)

37. X. Chai, Z. Gan, K. Yang, Y. Chen, X. Liu, An image encryption algorithm based on the memristive hyperchaotic system, cellular automata and DNA sequence operations. Signal Process. Image Commun. **52**, 6–19 (2017)

38. Y. Zhang, The image encryption algorithm based on chaos and DNA computing. Multimedia Tools Appl. **77**(16), 21589–21615 (2018)

39. L. Genelle, E. Prou_, M. Quisquater, Thwarting higher-order side channel analysis with additive and multiplicative maskings, in *Cryptographic Hardware and Embedded Systems – CHES 2011 – 13th In- Ternational Workshop, Nara, Japan, September 28 – October 1, 2011. Proceedings, Volume 6917 of Lecture Notes in Computer Science*, ed. by B. Preneel, T. Takagi, (Springer, 2011), pp. 240–255

40. B. Gierlichs, L. Batina, P. Tuyls, B. Preneel, Mutual information analysis, in *CHES, 10th International Workshop, Volume 5154 of Lecture Notes in Computer*, (Springer, Washington, DC, 2008), pp. 426–442

41. R. Gilmore, N. Hanley, M. O'Neill, Neural network based attack on a masked implementation of AES, in *Hardware Oriented Security and Trust (HOST), 2015 IEEE International Symposium*, (2015), pp. 106–111

42. S.O.A.F.M. Koko, A. Babiker, Comparison of various encryption algorithms and techniques for improving secured data communication. IOSR J. Comput. Eng. (IOSR-JCE) **17**(1), 62–69 (2015)

43. A. Kumar, D.S. Jakhar, M.S. Makkar, Comparative analysis between DES and RSA algorithm's. Int. J. Adv. Res. Comput. Sci. Soft. Eng. **2**(7), 386–391 (2012)

44. P. Mahajan, A. Sachdeva, A study of encryption algorithms AES, DES, and RSA for security. Global J. Comp. Sci. Technol. **13** (2013). https://computerresearch.org/index.php/computer/article/view/272/272

45. https://www.howtogeek.com/167783/htg-explains-the-difference-between-wep-wpa-and-wpa2-wireless-encryption-and-why-it-matters/

46. https://www.forbes.com/sites/forbestechcouncil/2019/09/23/why-5g-can-be-more-secure-than-4g/#26d12e8957b2

47. A.N. Bikos, N. Sklavos, LTE/SAE security issues on 4G wireless networks. IEEE Sec. Priv. **11**(2), 55–62 (2013). https://doi.org/10.1109/MSP.2012.136

Chapter 3
Cryptography Concepts: Integrity, Authentication, Availability, Access Control, and Non-repudiation

3.1 Integrity: Hashing Concept

To validate the integrity of the data transmitted over the channel, message authentication code (MAC) is used for checking the messages and the authentication, ensuring that the integrity of the information has not been modified under the transmission.

Hash functions are mathematical functions which map values in the domain values in the range. Hash functions are special mathematical functions that satisfy the following three properties:

- Inputs can be any size.
- Outputs are fixed.
- Efficiently computable, i.e., the mapping should be efficiently computable.

The purpose of a hash function is to produce a "fingerprint" of a message or data for authentication. A hash function $H(m)$ maps an input message m to a hash value h as described by eq. (3.1). Message m is of any arbitrary length. Hash h is fixed length. Often, h is called as the *message digest* of m. Figure 3.1 shows the abstract level of hashing functionalities:

$$h = H(m) \tag{3.1}$$

Cryptographic hash functions are one-way function as finding hash h from m is easy, but not vice-versa. Given a message, it is hard to find another message that has the same hash value. Given a hash function, it is hard to find two messages with the same hash value. In cryptography, the avalanche effect is a term associated with a specific behavior of mathematical functions used for encryption.

© The Editor(s) (if applicable) and The Author(s), under exclusive license to
Springer Nature Switzerland AG 2020
K. S. Mohamed, *New Frontiers in Cryptography*,
https://doi.org/10.1007/978-3-030-58996-7_3

Fig. 3.1 Hashing

3.1.1   SHA-3

SHA was designed by NIST and NSA in 1993, revised 1995 as **SHA-1**. The algorithm produces 160 bit hash values. Now, it is the most preferred hash algorithm. **SHA-2** uses the Davies–Meyer structure, an instance of the Merkle–Damgård structure, with a block cipher (sometimes called SHACAL-2) built out of an ARX network, like MD4. **SHA-3** (Secure Hash Algorithm 3) is the latest member of the Secure Hash Algorithm family of standards. SHA-3 uses the sponge construction in which is "absorbed" into the sponge, and then the result is "squeezed" out. In the absorbing phase, message blocks are XORed into a subset of the state, which is then transformed as a whole using a permutation function. In the "squeeze" phase, output blocks are read from the same subset of the state, alternated with the state transformation function. The size of the part of the state that is written and read is called the rate, and the size of the part that is untouched by input/output is called the capacity. The capacity determines the security of the scheme. The maximum security level is half the capacity. To ensure the message can be evenly divided into r-bit blocks, padding is required. Changing a single bit causes each bit in the output to change with 50% probability, demonstrating an **avalanche effect** [1, 2]. In a nutshell, SHA2 (and SHA1) are built using the Merkle–Damgård structure. SHA3 on the other hand is built using a sponge function.

3.1.1.1   Merkle–Damgård Construction

In cryptography, the Merkle–Damgård construction or Merkle–Damgård hash function is a method of building collision-resistant cryptographic hash functions from collision-resistant one-way compression functions. This construction was used in the design of many popular hash algorithms such as MD5, SHA1, and SHA2. The Merkle–Damgård hash function first applies an MD-compliant padding function to create an input whose size is a multiple of a fixed number (e.g., 512 or 1024) – this is because compression functions cannot handle inputs of arbitrary size. The hash function then breaks the result into blocks of fixed size and processes them one at a time with the compression function, each time combining a block of the input with the output of the previous round. In order to make the construction secure, Merkle and Damgård proposed that messages be padded with a padding that encodes the length of the original message [3–5].

3.1.1.2 Sponge Function

In cryptography, a sponge function or sponge construction is any of a class of algorithms with finite internal state that take an input bit stream of any length and produce an output bit stream of any desired length. Sponge functions have both theoretical and practical uses. They can be used to model or implement many cryptographic primitives, including cryptographic hashes, message authentication codes, mask generation functions, stream ciphers, pseudorandom number generators, and authenticated encryption. A sponge function is built from three components: a state memory, a function that transforms the state memory, and a padding function. The state memory is divided into two sections: the bitrate and the capacity. The padding function appends enough bits to the input string so that the length of the padded input is a whole multiple of the bitrate [6].

The sponge function operates as follows:

- The state S is initialized to zero.
- The input string is padded.
- For each r-bit block B of the padded input:

  - R is replaced with R XOR B
  - S is replaced by $f(S)$

- The R portion of the state memory is output.
- repeat until enough bits are output:

  - S is replaced by $f(S)$
  - The R portion of the state memory is output

If less than r bits remain to be output, then R will be truncated.

3.1.2 HMAC-SHA256 (Hashed Message Authentication Code, Secure Hash Algorithm)

3.1.2.1 Nomenclature

B	Block size (in bytes) of the input to the approved hash function.
H	SHA-256 hash function.
ipad	Inner pad; the byte x'36' repeated B times.
K	Secret key shared between the host and the device.
K_0	The key K after any necessary pre-processing to form a B-byte key.
L	Block size (in bytes) of the output of the approved hash function.
opad	Outer pad; the byte x'5c' repeated B times.
t	The number of bytes of MAC.
text	The message on which the HMAC is calculated; text does not include the padded key.

x′N′   Hexadecimal notation, where each symbol in the string "N" represents four binary bits.

‖   Concatenation.

xor   Exclusive-OR operation.

a, b, c   Working variables that are the w-bit words used in the computation of the hash values, $H^{(i)}$.

$H^{(i)}$ The ith hash value. $H^{(0)}$ is the initial hash value; $H^{(N)}$ is the final hash value and is used to determine the message digest.

$H_j^{(i)}$ The jth word of the ith hash value, where $H_0^{(i)}$ is the left-most word of hash value i.

K_t Constant value to be used for iteration t of the hash computation.

k Number of zeroes appended to a message during the padding step.

l Length of the message, M, in bits.

m Number of bits in a message block, $M^{(i)}$.

M Message to be hashed.

$M^{(i)}$ Message block i, with a size of m bits.

$M_j^{(i)}$ The jth word of the ith message block, where $M_0^{(i)}$ is the left-most word of message block i

n Number of bits to be rotated or shifted when a word is operated upon.

N Number of blocks in the padded message.

T Temporary w-bit word used in the hash computation.

w Number of bits in a word.

W_t The tth w-bit word of the message schedule.

^   Bitwise AND operation.

Or   Bitwise OR ("inclusive-OR") operation.

Xor   Bitwise XOR ("Exclusive-OR") operation.

!   Bitwise complement operation.

+   Addition modulo 2^w.

≪   Left-shift operation, where $x \ll n$ is obtained by discarding the left-most n bits of the word x and then padding the result with n zeroes on the right.

≫   Right-shift operation, where $x \gg n$ is obtained by discarding the rightmost n bits of the word x and then padding the result with n zeroes on the left.

3.1.2.2   Introduction

HMAC_SHA is a secure cryptographic hash function-based message and shared key authentication protocols. It can effectively prevent data being intercepted and tampered with during transmissions.

The message authentication code (MAC) is calculated using HMAC SHA-256 as defined in [7–9]. The HMAC SHA-256 calculation takes as input a key and a message. The resulting MAC is 256 bits (32 bytes), which are embedded in the data frame as part of the request or response. The key used for the MAC calculation is 256 bit authentication key. The receiver computes the MAC on the received data using the same key and HMAC function as was used by the transmitter and compares the result computed with the received MAC. If the two values match, then the data has been correctly received.

The HMAC-SHA-256 is the implementation of the HMAC message authentication code calculation using the SHA-256 hash function. The hash function can be iterative and breaks up the message into blocks of a fixed size and iterates over them. For SHA-256, the message size is less than 2^{64}. Each message block has 512 bits. The message is padded so that it can be parsed into N x 512 blocks. The output of HMAC calculation is the same as the underlying hash function which in our case is 256 bits (32 bytes) wide. In this paper, we present a low latency hardware implementation of HMAC-SHA-256(Hashed Message Authentication Code, Secure Hash Algorithm) algorithm for Internet of Things (Io) applications. Figure 3.2 and Table 3.1 illustrates the step-by-step process in the HMAC algorithm. SHA-256 is SHA-2 with 256-bit output.

3.1.2.3 HMAC Algorithm

As specified in [7], the size of a message block for Secure Hash Algorithm SHA-256 is 512 bits (i.e., $B = 64$ bytes). To compute a MAC over the data "text" using the HMAC function, the following operation is performed:

$$MAC(\text{text})t = HMAC(K, \text{text})t$$
$$= H\big((K0 \text{ xor opad}) \| H\big((K0 \text{ xor ipad}) \| \text{text}\big)\big)t \qquad (3.2)$$

3.1.2.4 SHA-256 Algorithm

1. Pad the message, M:

The message, M, shall be padded before hash computation begins. The purpose of this padding is to ensure that the padded message is a multiple of 512 or 1024 bits, depending on the algorithm. Suppose that the length of the message, M, is l bits. Append the bit "1" to the end of the message, followed by k zero bits, where k is the smallest, nonnegative solution to the equation $1 + 1 + k = 448 \bmod 512$. Then append the 64 bit block that is equal to the number 1 expressed using a binary representation. For example, the (8 bit ASCII) message "abc" has length $8 \times 3 = 24$, so the message is padded with a one bit, then $448 - (24 + 1) = 423$ zero bits, and then the message length, to become the 512-bit padded message as shown in Fig. 3.3.

2. Parse the padded message into N 512-bit message blocks, $M^{(1)}, M^{(2)}, \ldots, M^{(N)}$.

3. Set the initial hash value, $H^{(0)}$:

Before hash computation begins for each of the secure hash algorithms, the initial hash value, $H^{(0)}$, must be set. The size and number of words in $H^{(0)}$ depends on the message digest size.

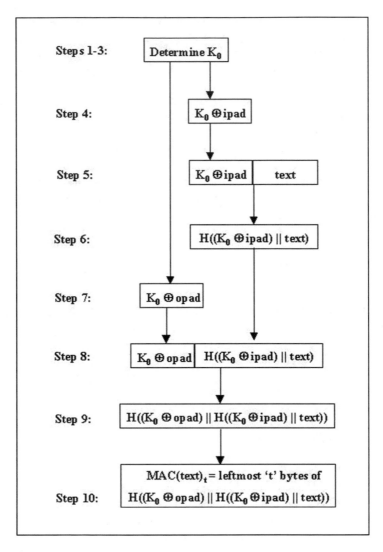

Fig. 3.2 Illustration of the HMAC construction

For SHA-256, the initial hash value, $H^{(0)}$, shall consist of the following eight 32 bit words, in hex:

$$H_0^{(0)} = 6a09e667$$
$$H_1^{(0)} = bb67ae85$$
$$H_2^{(0)} = 3c6ef372$$
$$H_3^{(0)} = a54ff53a$$

Table 3.1 General HMAC algorithm

Steps	Step-by-step description
Step 1	K = 32 bytes, B = 64 bytes, $K \neq B$
Step 2	$K < B$: append zeros to the end of K to create a B-byte string K_0 (K will be appended with 32 zero bytes 0x00)
Step 3	Exclusive-OR $K0$ with *ipad* to produce a B-byte string: K_0 xor *ipad*
Step 4	Append the stream of data *text* to the string resulting from step 4: (K_0 xor *ipad*) ‖ *text*
Step 5	Apply H to the stream generated in step 5: **H((K_0 xor *ipad*) ‖ *text*)**
Step 6	Exclusive-OR K_0 with *opad*: K_0 xor *opad*
Step 7	Append the result from step 5 to step 6: (K_0 xor *opad*) ‖ **H((K_0 xor *ipad*) ‖ *text*)**
Step 8	Apply H to the result from step 7: **H((K_0 xor *opad*) ‖ H((K_0 xor *ipad*) ‖ *text*))**
Step 9	Select the leftmost t bytes of the result of step 8 as the MAC

Fig. 3.3 SHA-256 message example

$H_4^{(0)} = 510e527f$

$H_5^{(0)} = 9b05688c$

$H_6^{(0)} = 1f83d9ab$

$H_7^{(0)} = 5be0cd19$

4. After preprocessing is completed, each message block, $M^{(1)}$, $M^{(2)}$, ..., $M^{(N)}$, is processed in order, using the following steps:
   For $i = 1$ to N:
{

1. Prepare the message schedule, $\{W_t\}$

$$W_t = \begin{cases} M_t^{(t)} & 0 \leq t \leq 15 \\ \sigma_1^{\{256\}}\left(W_{t-2}\right) + W_{t-7} + \sigma_0^{\{256\}}\left(W_{t-15}\right) + W_{t-16} & 16 \leq t \leq 63 \end{cases}$$

2. Initialize the eight working variables, a, b, c, d, e, f, g, and h, with the $(i-1)$st hash value:

$$a = H_0^{(i-1)}$$
$$b = H_1^{(i-1)}$$
$$c = H_2^{(i-1)}$$
$$d = H_3^{(i-1)}$$
$$e = H_4^{(i-1)}$$
$$f = H_5^{(i-1)}$$
$$g = H_6^{(i-1)}$$
$$h = H_7^{(i-1)}$$

3. For $t = 0$ to 63

{

$$T_1 = h + \sum_1^{\{256\}}(e) + Ch(e,f,g) + K_t^{\{256\}} + W_t$$

$$T_2 = \sum_0^{\{256\}}(a) + Maj(a,b,c)$$

$$h = g$$

$$g = f$$

$$f = e$$

$$e = d + T_1$$

$$d = c$$

$$c = b$$

$$b = a$$

$$a = T_1 + T_2$$

}

4. Compute the ith intermediate hash value $H^{(i)}$:

$$H_0^{(i)} = a + H_0^{(i-1)}$$
$$H_1^{(i)} = b + H_1^{(i-1)}$$
$$H_2^{(i)} = c + H_2^{(i-1)}$$

$$H_3^{(i)} = d + H_3^{(i-1)}$$
$$H_4^{(i)} = e + H_4^{(i-1)}$$
$$H_5^{(i)} = f + H_5^{(i-1)}$$
$$H_6^{(i)} = g + H_6^{(i-1)}$$
$$H_7^{(i)} = h + H_7^{(i-1)}$$
}

After repeating steps one through four a total of N times (i.e., after processing $M^{(N)}$), the resulting 512 bit message digest of the message, M, is

$$H_0^{(N)} H_1^{(N)} H_2^{(N)} H_3^{(N)} H_4^{(N)} \Big| H_5^{(N)} H_6^{(N)} \Big| H_7^{(N)}.$$

3.1.2.5   HMAC-SHA256 Illustrative Example

This example for the HMAC-SHA256 calculation uses one message of 284 bytes and a key of 32 bytes.

Input:

Text: 284 bytes message

> 000102030405060708090a0b0c0d0e0f101112131415161718191a1b1c1d1
> e1f202122232425262728292a2b2c2d2e2f303132333435363738393a3b3c3
> d3e3f404142434445464748494a4b4c4d4e4f50515253545556575859a5b
> 5c5d5e5f606162636465666768696a6b6c6d6e6f707172737475767
> 778797a7b7c7d7e7f808182838485868788898a8b8c8d8e8f9091929394
> 95969798999a9b9c9d9e9fa0a1a2a3a4a5a6a7a8a9aaabacadae
> afb0b1b2b3b4b5b6b7b8b9babbbcbdbebfc0c1c2c3c4c5c6c7c8c9cacbcccd-
> cecfd0d1d2d3d4d5d6d7d8d9dadbdcdddedfe0e1e2e3e4e5e6e7e8e9eaebeced
> eeeff0f1f2f3f4f5f6f7f8f9fafbfcfdfeff00000000000000000000000
> 0000000000000000010000000100000003.

Key: 32 bytes.

> 0000000000000000000000000000000000000000001101011011001010101
> 111001

Steps:

1. K_0: append 32 bytes of zeros to the end of the key to create a 64 byte message

> 0000000000000000000000000000000000000000001101011011001010101111
> 0010000000000000000000000000000000000000000000000000000000000000
> 000000

2. Create a 64 byte ipad pattern of repeating the hexadecimal value 0x36.

   3636363636363636363636363636363636363636363636363636363636363636
   3636363636363636363636363636363636363636363636363636363636363636
   363636

3. Create a 64 byte opad pattern of repeating the hexadecimal value 0x5c.

   5c5c5c5c5c5c5c5c5c5c5c5c5c5c5c5c5c5c5c5c5c5c5c5c5c5c5c5c5c5c5c5c5c
   5c5c5c5c5c5c5c5c5c5c5c5c5c5c5c5c5c5c5c5c5c5c5c5c5c5c5c5c5c5c5c5c5c

4. XOR K_0 with ipad: K_0 xor *ipad*.

   3636363636363636363636363636363636363637262627372637373727
   2637363636363636363636363636363636363636363636363636363636363636
   363636

5. Concatenate text at the end: $(K_0$ xor *ipad*$)$ ∥ *text*

   3636363636363636363636363636363636363637262627372637373727
   2637363636363636363636363636363636363636363636363636363636363636
   3636360001020304050607080900a0b0c0d0e0f10111213141516171819191a1b
   1c1d1e1f202122232425262728292a2b2c2d2e2f303132333435363738393a
   3b3c3d3e3f404142434445464748494a4b4c4d4e4f50515253545556575859
   5a5b5c5d5e5f606162636465666768696a6b6c6d6e6f707172737475767
   778797a7b7c7d7e7f80818283848586878889 8a8b8c8d8e8f909192939495
   969798999a9b9c9d9e9fa0a1a2a3a4a5a6a7a8a9aaabacadaeafb0
   b1b2b3b4b5b6b7b8b9babbbcbdbebfc0c1c2c3c4c5c6c7c8c9cacbcccd-
   cecfd0d1d2d3d4d5d6d7d8d9dadbdcdddedfe0e1e2e3e4e5e6e7e8e9eaebeced
   eeeff0f1f2f3f4f5f6f7f8f9fafbfcfdfeff000000000000000000000000000000
   00000000010000000100000003

6. Hash the value above using SHA-256 algorithm.

   The message is extended as shown above. In this case the data message to be
   hashed is 2784 bits (348 bytes). To achieve a padded message that is a multiple of 512 bits, the 2784 bits message is padded with one bit, then 223 zero
   bits, and then 64 bits representing the message length. The message now
   becomes a 6*512 bit padded message. This padded message is then processed using the SHA-256 algorithm explained above to obtain the
   hash value.
   663cb567af6aeb40a9948a1621c129b69ffbd41244596df0b2729bac5ac0fbb1

7. XOR appended key with opad: K_0 xor *opad*

   5c5c5c5c5c5c5c5c5c5c5c5c5c5c5c5c5c5c5c5c5d4c4c4d5d4c5d5d5d4d4c5d
   5c5c5c5c5c5c5c5c5c5c5c5c5c5c5c5c5c5c5c5c5c5c5c5c5c5c5c5c5c5c5c5c5c

8. Concatenate the result from step 6 and step 7:

   $(K_0$ xor *opad*$)$ ∥ $H((K_0$ xor *ipad*$)$ ∥ *text*$)$
   5c5c5c5c5c5c5c5c5c5c5c5c5c5c5c5c5c5c5c5c5d4c4c4d5d4c5d5d5d4d4c5
   d5c5c5c5c5c5c5c5c5c5c5c5c5c5c5c5c5c5c5c5c5c5c5c5c5c5c5c5c5c5c5c5c

5c663cb567af6aeb40a9948a1621c129b69ffbd41244596df0b2729bac5ac
0fbb1

9. Hash the output of step 8 using SHA-256

The message should be padded before hashing of SHA-256 algorithm. In this case the message length is 768 bits (96 bytes). To achieve a padded message that is a multiple of 512 bits, the message is padded with one bit and then 191 zero bits and then 64 bits representing the message length. The message is now 2*512 bit padded message. This padded message is then processed using the SHA-256 algorithm explained above to obtain the hash value.a771af5ea520f789c0e6668c4c9cb6f30111d381503917af47af37c5 3b7adca2.

3.1.3   MD5 Algorithm

MED5   algorithm works as follows:

- Pad message so its length is 448 mod 512.
- Append a 64 bit length value to message.
- Initialize four-word (128 bit) MD buffer (A, B, C, D).
- Process message in 16-word (512 bit) blocks: using four rounds of 16 bit operations on message block and buffer. Add output to buffer input to form new buffer value.
- Output hash value is the final buffer value.

3.1.4   Blake

BLAKE is a cryptographic hash function based on Dan Bernstein's ChaCha stream cipher, but a permuted copy of the input block, XORed with round constants, is added before each ChaCha round. Like SHA-2, there are two variants differing in the word size. ChaCha operates on a 4 × 4 array of words. BLAKE repeatedly combines an 8-word hash value with 16 message words, truncating the ChaCha result to obtain the next hash value. BLAKE-256 and BLAKE-224 use 32 bit words and produce digest sizes of 256 bits and 224 bits, respectively, while BLAKE-512 and BLAKE-384 use 64 bit words and produce digest sizes of 512 bits and 384 bits, respectively. The BLAKE2 hash function, based on BLAKE, was announced in 2012. The BLAKE3 hash function, based on BLAKE2, was announced in 2020. BLAKE2 is a cryptographic hash function **faster than MD5, SHA-1, SHA-2, and SHA-3** (Fig. 3.4), yet is at least as secure as the latest standard SHA-3 [11].

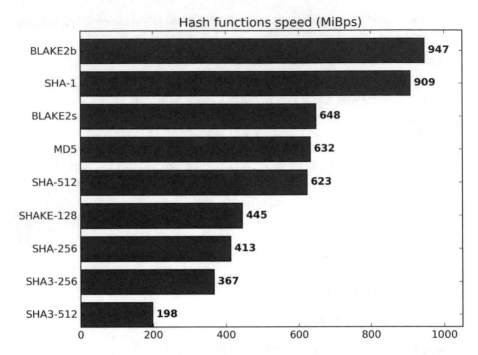

Fig. 3.4 Hash functions speed [10]

3.2   Integrity: Digital Signature

In digital signature, we reverse the role of public and private keys. An encrypted hash is called a "digital signature" as shown in Fig. 3.5. Digital signature is used I e-commerce. For the real system, we are sending the message with the digital signature as shown in Fig. 3.6.

3.3   Authentication

Authentication answers the following question "how does a receiver know that remote communicating entity is who it is claimed to be?". Authentication ensures that message has not been altered. Message is from alleged sender. Message sequence is unaltered. An example to show an authenticated email is shown in Fig. 3.7 where Alice wants to provide sender authentication message integrity, so Alice digitally signs message and sends both message and digital signature.

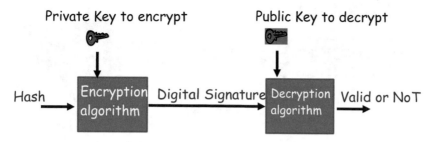

Fig. 3.5 Digital signature

Fig. 3.6 Sending the
message with the digital
signature

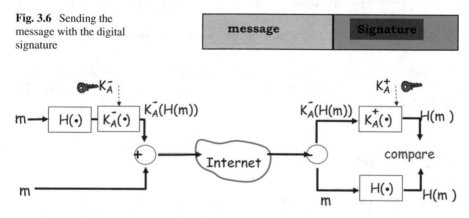

Fig. 3.7 Authenticated email

3.3.1   HDCP

HDCP is an acronym for High-Bandwidth Digital Content Protection. HDCP is a specification developed by Intel Corporation to protect digital entertainment content across the DVI/HDMI interface.

Digital Content Protection LLC (DCP) is the organization that licenses HDCP. How HDCP works is explained in Fig. 3.8. Each HDCP transmitter and receiver has 40 unique 56 bit private keys. These keys are provided by DCP to licensed HDCP chip vendors, who preload the keys onto the chips before selling them to device manufacturers. These keys must never leave the chip and may not be read by any other device.

Each HDCP chip also has a public 40 bit value known as the Key Selection Vector (KSV). Each KSV consists of 20 binary 1 s and 20 binary 0 s. The KSVs and keys of all licensed HDCP devices are mathematically related according to a cryptographic key exchange protocol. In this scheme, any two licensed devices

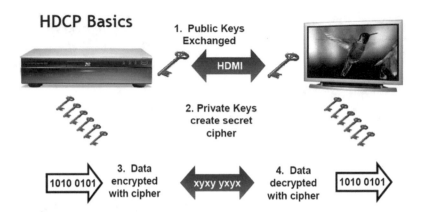

Fig. 3.8 How HDCP works. **It consists of authentication, data encryption**

can swap KSVs and use them, along with their private keys, to come up with a shared secret key. This shared key can be used to encrypt and decrypt the TMDS stream.

3.3.2 CAPTCHA Codes

CAPTCHA is a verification process that requires users to enter a predetermined code. CAPTCHA stands for *Completely Automated Public Turing test to tell Computers and Humans Apart*. CAPTCHA exists to prevent spam from automated form submissions that can fill your site with junk postings, spam user accounts, or worse feel for security holes to be exploited on your website. CAPTCHAs work by providing a question that is simple for a human to answer but difficult for a bot to answer. Different types of CAPTCHAs have been developed, including simple math questions and classification problems [12].

3.4　Availability: Intrusion Detection

3.4.1　Artificial Immune System

Any human are exposed to a huge range of harmful microorganisms (pathogens or antigens) which can be either single cell or multicellular such as bacteria, parasites, and viruses. These microorganisms can damage the human body. The human natural immune system (NIS) is a very complex defense system that can prevent this damage. There is no central organ controlling the functionality of the immune system. The parts of the NIS are:

1. Blood: white blood cells in particular.
2. Lymph nodes: stores T and B cells and **traps** antigens/pathogens.
3. Thymus Gland: produces T Lymphocytes that circulating throughout the body looking for abnormalities.
4. Bone Marrow: produces B lymphocytes.

Immune organs as positioned throughout the body are shown in Fig. 3.9.

The immune system has a learning phase over the years. Every element that can be recognized by the immune system is called an *antigen.* The cells that originally belong to our body and are harmless to its functionality are termed *(self-antigens)*, while the disease causing elements are named *(non-self-antigens)*. As the body is attacked by certain microorganisms, the immune system memorizes them to be able to differentiate between foreign cells and the body cells.

The immune system is composed of two different cells which are called B cell and T cell. These two types of cells are rather similar but differ with relation to how they recognize antigens and by their functional roles. B cells can recognize antigens in the blood stream, while T cells require antigens to be presented by other accessory cells. Antigens are covered with molecules, named *epitopes*. These allow them to be recognized by the receptor molecules on the surface of B cells, called *antibodies (Ab)*.

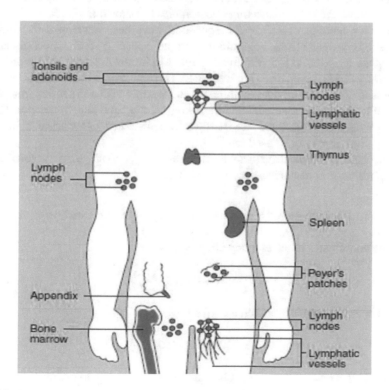

Fig. 3.9 Immune organs are positioned throughout the body

NIS is multilayered system. The human immune system (HIS) consists of three levels: skin, innate immunity, and adaptive immunity. The skin is the first barrier. The second barrier is physiological conditions such as temperature which provides inappropriate living conditions for harmful organisms. Once pathogens have entered the body, they are dealt with by the innate such as phagocytes which are cells that protect the body by ingesting harmful foreign particles and by the acquired immune response system (lymphocyte) which is considered as a detector.

Pathogens are detected when a molecular bond is established between the pathogen and receptors that cover the surface of the lymphocyte. The number of receptors that bind to pathogens will determine the affinity that the lymphocyte has for a given pathogen. If a bond is very likely to occur, then many receptors will bind to pathogen epitopes, resulting in a high affinity for that pathogen; if a bond is unlikely to occur, then few receptors will bind to epitopes, and the lymphocyte will have a low affinity for that pathogen. Lymphocytes can only be activated by a pathogen if the lymphocyte's affinity for the pathogen exceeds a certain affinity threshold. The layers are shown in Fig. 3.10. How B cells recognizes an antigen is shown in Fig. 3.11 and an illustrative example is shown in Fig. 3.12. Moreover, Fig. 3.13 shows the binding process.

In the 1990s, **artificial immune system (AIS)** that is inspired by the natural or biological immune system has emerged as a new branch in computational intelligence systems. AIS as a connecting two domains are shown in Fig. 3.14. Until now a significant number of AIS-based algorithms have been developed for different applications such as pattern recognition, computer security, fault detection, and so many other applications [13–30]. In this paper and for the first time, AIS is proposed for security at RTL level. The proposed system is implemented by Verilog and tested on Xilinx FPGA as it provides an efficient and flexible platform for fast implementation of hardware architecture and fast reprogramming and experimental testing of various revised versions of the same hardware architectures. Other artificial machine learning algorithms can be used too [31].

Three basic concepts in artificial immune system (AIS) are affinity, clonal selection algorithm, and negative selection algorithm.

A. Affinity

In AIS, the affinity between antibodies and antigens is calculated using Euclidean distance.

If the coordinates of an antibody are given by:

$$Ab = \left(Ab_1, Ab_2, \ldots, Ab_n\right) \tag{3.3}$$

and the coordinates of an antigen are given by:

$$Ag = \left(Ag_1, Ag_2, \ldots, Ag_n\right) \tag{3.4}$$

then the distance between them is obtained by the following equation:

THE IMMUNE SYSTEM

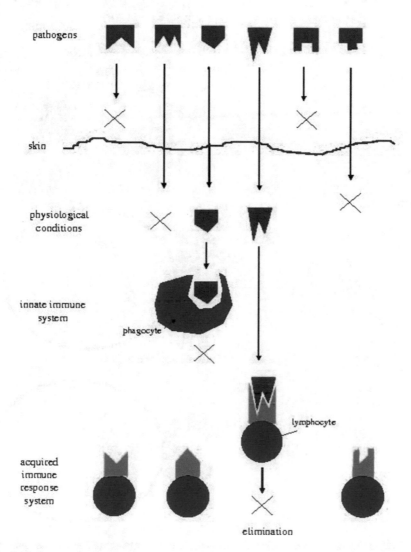

Fig. 3.10 Layers of protection in the human body

$$D = \sqrt{\sum_{i=o}^{n} \left(Ab_i - Ag_i \right)^2} \qquad (3.5)$$

Fig. 3.11 B cell
recognizes an antigen

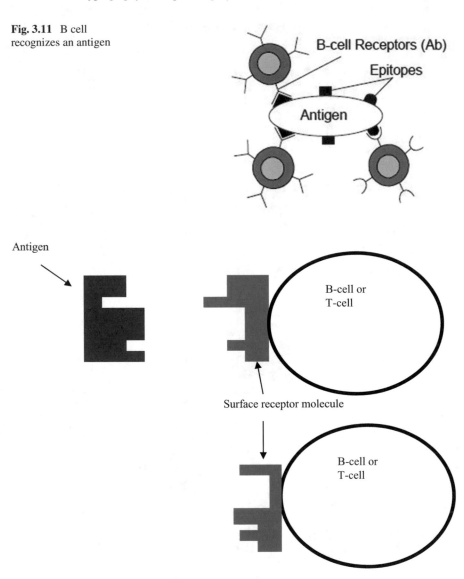

Fig. 3.12 (**a**) Lymphocyte recognizes the red pathogen, (**b**) lymphocyte does not recognize the red pathogen

B. Negative Selection Algorithm and Clonal Selection
The negative selection algorithm is the main algorithm used for NIS. P is the self set. C is the generated candidate to be non-self. M is the detector. P^* is the patterns to be protected (suspicion). The algorithm can be summarized as follows (Fig. 3.15):

1. Generate random candidate elements (C).

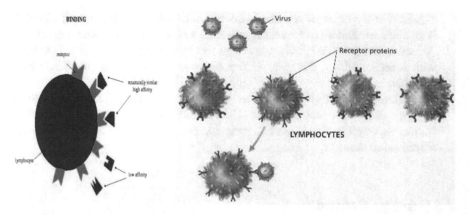

Fig. 3.13 Pathogens are detected when a molecular bond is established between the pathogen and receptors that cover the surface of the lymphocyte

Fig. 3.14 AIS is connecting two domains

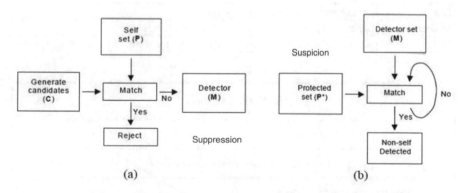

Fig. 3.15 The **negative selection algorithm is the main algorithm used for NIS.** (**a**) Generating the set of detectors. (**b**) Monitoring for the presence of undesired (non-self) patterns. P is the self set. C is the generated candidate to be non-self. M is the detector. $P*$ is the patterns to be protected (suspicion)

2. Compare the elements in C with the elements in P. If a match occurs, then reject it. It is a self. Else store this element of C in the detector set M.
3. After generating the set of detectors (M), the next stage of the algorithm consists in monitoring the system for the presence of non-self patterns. For all

elements of the detector set, that corresponds to the non-self patterns, check if
it matches an element of $P*$ and, if yes, then a non-self pattern was recognized.
4. An action has to be taken. When a B-cell receptor recognizes a non-self antigen
   with a certain affinity, it is selected to proliferate and produce antibodies in high
   volumes.
5. The activated B cells with high antigenic affinities are selected to become
   memory cells with long life spans. These memory cells are pre-eminent in
   future responses to this same antigenic pattern, or a similar one this is called
   clonal selection.

3.5   Access Control

Security levels and solutions are summarized in Table 3.2 and Fig. 3.16. Passwords
are shared secret between two parties. Smart cards are electronics embedded in card
capable of providing long passwords or satisfying challenge. It may have display to
allow reading of password or can be plugged in directly. Biometrics uses of one or
more intrinsic physical or behavioral traits to identify someone such as fingerprint
reader, palm reader, retinal scan, or biometrics in general. A VPN, or virtual private

Table 3.2 Security levels and solutions

Security level	Solution
Low	Something you know (pin, password)
Medium	Something you know + something you have (access card)
High	Something you know + something you have + something you are (fingerprint)

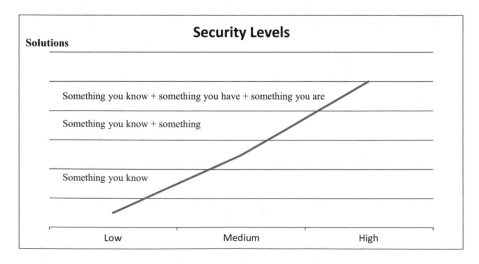

Fig. 3.16 Access control: security levels

network, allows you to create a secure connection to another network over the Internet. VPNs can be used to access region-restricted websites, shield your browsing activity from prying eyes on public Wi-Fi, and more. When you connect your computer (or another device, such as a smartphone or tablet) to a VPN, the computer acts as if it's on the same local network as the VPN. All your network traffic is sent over a secure connection to the VPN. Because your computer behaves as if it's on the network, this allows you to securely access local network resources even when you're on the other side of the world. You'll also be able to use the Internet as if you were present at the VPN's location, which has some benefits if you're using pubic Wi-Fi or want to access geo-blocked websites [32].

3.6 Non-repudiation: Trusted Third Party

The ability to ensure that a party to a contract or a communication cannot deny the authenticity of their signature on a document or the sending of a message that they originated. You can't deny doing something you did. To mitigate the risk of people repudiating their own signatures, the standard approach is to involve a trusted third party. For digital information, the most commonly employed **TTP** is a certificate authority, which issues public key certificates. A public key certificate can be used by anyone to verify digital signatures without a shared secret between the signer and the verifier. The role of the certificate authority is to authoritatively state to whom the certificate belongs, meaning that this person or entity possesses the corresponding private key. However, a digital signature is forensically identical in both legitimate and forged uses. Someone who possesses the private key can create a valid digital signature [33].

3.7 Conclusions

This chapter explores different cryptography concepts such as authentication, integrity, availability, access control, and non-repudiation. It presents concepts of digital signatures, hash functions, and message authentication codes (MACs).

References

1. J. Kelsey, *SHA3, Past, Present, and Future* (CHES, 2013)
2. M.J. Dworkin, SHA-3 Standard: Permutation-Based Hash and Extendable-Output Functions. Federal Inf. Process, STDS, (NIST FIPS) – 202, (2015)
3. Menezes, van Oorschot, Vanstone, Handbook of Applied Cryptography, Chapter 9, (2001)
4. J. Katz, Y. Lindell, *Introduction to Modern Cryptography* (Chapman and Hall/CRC Press, 2007), p. 134, (construction 4.13),

5. M. Nandi, S. Paul, Speeding up the widepipe: secure and fast hashing, in *Indocrypt 2010*, ed. by G. Gong, K. Gupta, (Springer, 2010)
6. G. Bertoni, J. Daemen, M. Peeters, A. Gilles Van, *On the Indifferentiability of the Sponge Construction* (EuroCrypt, 2008)
7. [HMAC-SHA], D. Eastlake, T. Hansen, *US Secure Hash Algorithms (SHA and MAC-SHA)* (RFC 4634, 2006)
8. National Institute of Standards and Technology, Secure Hash Standard (SHS), Federal Information processing Standards Publication 180-1, 17 April 1995
9. National Institute of Standards and Technology, The Keyed-Hash Message Authentication Code (HMAC) Federal Information processing Standards Publication 198, 6 March 2002
10. https://blake2.net/
11. https://en.wikipedia.org/wiki/BLAKE_(hash_function)
12. https://internet.com/website-building/how-to-add-a-captcha-to-your-website/
13. S.J. Nanda, G. Panda, B. Mekhi, Improved identification of nonlinear dynamic systems using artificial immune system, in *IEEE International Conference on Control, Communication and Automation (INDICON-08)*, (IIT Kanpur, 2008), pp. 268–273
14. S. Forrest, S. Hofmeyr, A. Somayaji, T.A. Longstaff, A sense of self for Unix processes, in *Proceedings of the IEEE Symposium on Computer Security and Privacy*, (1996), pp. 120–128
15. A. Somayaji, S. Hofmeyr, S. Forrest, Principle of a computer immune system, in *Proceeding of the Second New Security Paradigms Workshop*, (1997), pp. 75–82
16. S. Hofmeyr, S. Forrest, A. Somayaji, Intusion detection using sequences of system calls. J. Comput. Secur. **6**, 151–180 (1998)
17. U. Aickelin, J. Greensmith, J. Twycross, Immune system approaches to intrusion detection – A review, in *Proceeding of the Third International Conference on Artificial Immune Systems, ICARIS-04*, (2004), pp. 316–329
18. L.N. de Charsto, J.V. Zuben, Learning and optimization using clonal selection principle. IEEE Trans Evol. Comput. Spec. Issue Artific. Immun. Syst. **6**(3), 239–251 (2002)
19. L.N. de Charsto, J. Timmis, An artificial immune network for multimodal function optimization, in *IEEE Congress on Evolutionary Computation (CEC'02)*, vol. 1, (Hawaii, 2002), pp. 699–674
20. A.B. Watkins, Exploiting Immunological Metaphors in the Development of Serial, Parallel, and Distributed Learning Algorithms, (Ph. D thesis, University of Kent, 2005)
21. P.K. Harmer, P.D. Williams, G.H. Gunsch, G.B. Lamont, An artificial immune system architecture for computer security applications. IEEE Trans. Evol. Comput. **6**(3), 252–280 (2002)
22. Z. Zhang, T. Xin, Immune algorithm with adaptive sampling in noisy environments and its application to stochastic optimization problems. IEEE Comput. Intell. Mag., 29–40 (2007)
23. M. Drozda, S. Schaust, H. Szczerbicka, AIS for misbehavior detection in wireless sensor networks: Performance and design principles. IEEE Cong. Evol. Comput. (CEC'07), 3719–3726 (2007)
24. H.-W. Ge, L. Sun, Y.C. Liang, F. Qian, An effective PSO and AIS-based hybrid intelligent algorithm for job shop scheduling. IEEE Trans. Syst. Man Cybern. A **38**(2), 358–368 (2008)
25. S. Cayzer, U. Aickelin, A recommender system based on Idiotypic artificial immune networks. J. Math. Modell. Algor. **4**(2), 181–198 (2005)
26. Q. Chen, U. Aickelin, Movie recommendation systems using an artificial immune system, in *Proceedings of ACDM −04*, (Bristol, UK, 2004)
27. E. Alizadeh, N. Meskin, K. Khorasani, A negative selection immune system inspired methodology for fault diagnosis of wind turbines. IEEE Trans. Cybernet. **47**(11) (2017)
28. D. Dasgupta, Advances in artificial immune systems. IEEE Comput. Intell. Mag. **1**(4), 40–49 (2006)
29. W. Zhang, G.G. Yen, Z. He, Constrained optimization via artificial immune system. IEEE Trans. Cybern. **44**(2), 185–198 (2014)

30. Y. Ding, An immune system-inspired reconfigurable controller. IEEE Trans. Control Syst. Technol. **24**(5) (2016)
31. K.S. Mohamed, *Machine Learning for Model Order Reduction*, vol 664 (Springer, 2018)
32. https://www.howtogeek.com/133680/htg-explains-what-is-a-vpn/
33. https://www.sciencedirect.com/topics/computer-science/nonrepudiation

Chapter 4
New Trends in Cryptography: Quantum, Blockchain, Lightweight, Chaotic, and DNA Cryptography

4.1 DNA Cryptography

DNA cryptography is a promising and rapid emerging field in data security. DNA cryptography may bring forward a new hope for unbreakable algorithms. DNA cryptology combines cryptology and modern biotechnology. To encrypt using DNA, sender generates a DNA encoding table, and receiver generates another table through the same encoding technique and sends a clue to the sender to be able to generate it locally. The plaintext to be encoded is divided into two halves equally. If the plaintext is not even, we insert random padding. One half of the plaintext is converted into DNA sequence using sender-based table, and the other half of the plaintext is converted into DNA sequence using receiver-based table. DNA cryptography is a bio-inspired novel technique used for securing end to end communication, where DNA is used as an information carrier. DNA cryptography is assumed to be unbreakable algorithm [23–26]. The advantages of DNA computing over traditional computing are as follows [27]:

- *Speed:* Conventional computers have been known to perform approximately 10^8 instructions per second (MIPS). Combining DNA strands has been predicted to make computations equivalent to 10^9.
- *Storage:* DNA stores memory at the rate of 1 bit/nm^3, whereas conventional storage media can store 1 bit/10^{12} nm^3.
- *Power Requirements:* DNA computing does not require power, while computation is taking place. The chemical reactions that create the building blocks of DNA take place without any external power source.

In a nutshell, DNA computing has the characteristics of high parallelism, large storage capacity, and low-energy consumption [33–35]. DNA cryptography has a wide range of applications and can be implemented in various fields like mobile networks, cloud computing, IoT devices, real-time applications, the Internet, and multicast applications to secure plaintext messages, images, videos, servers, etc. [36, 37].

© The Editor(s) (if applicable) and The Author(s), under exclusive license to
Springer Nature Switzerland AG 2020
K. S. Mohamed, *New Frontiers in Cryptography*,
https://doi.org/10.1007/978-3-030-58996-7_4

4.1.1   Fundamentals of DNA Computing

DNA means deoxyribonucleic acid formed using four basic nucleic acids, namely, adenine (A), cytosine (C), guanine (G), and thymine (T) as depicted in Fig. 4.1. The pairs as (A, T) and (C, G) complement each other. These alphabets can be easily assigned to binary values *(A-00, C-01, G-10, T-11)*. By these encoding rules, there are 4! = 24 possible encoding methods. However, only eight coding combinations are suitable for the principle of complementarity. Because the binary numbers "0" and "1" are complementary, "00" and "11" and "01" and "10" are also complementary [23, 24].

4.1.2   DNA Cryptography Algorithm

DNA encryption can be performed through the following steps:

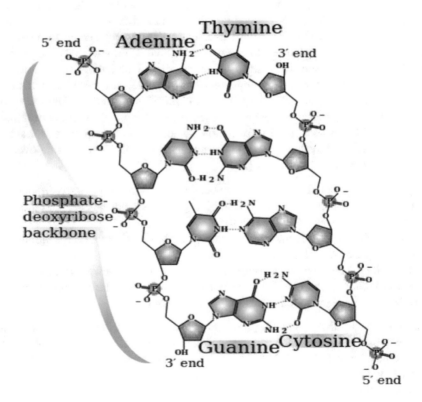

Fig. 4.1 Bases of DNA block

S	DNAseq	S	DNAseq	S	DNAseq	S	DNAseq	S	DNAseq	S	DNAseq
' '	ACAT	0	CAAA	@	CCAC	P	TTCA	'	TCCG	p	GACA
!	AGGT	1	CACC	A	TACT	Q	TTTA	a	GAGC	q	GACT
"	AAAG	2	CCGT	B	TCCT	R	TAGA	b	GTGC	r	GGAT
#	AGAC	3	CGAG	C	TACG	S	TGAG	c	GACG	s	GGTG
$	AAGC	4	CCTT	D	TGCC	T	TAAA	d	GTAA	t	GCTT
%	AACT	5	CCGT	E	TCTA	U	TGAC	e	GTAC	u	GACC
&	AGAA	6	CTGT	F	TAGT	V	TGAG	f	GCCT	v	GACT
'	AATC	7	CTCT	G	TTAA	W	TAAC	g	GCTA	w	GCCC
(	ATTG	8	CCGT	H	TGGC	X	TCCT	h	GAGT	x	GATC
)	AATT	9	CTCA	I	TGTT	Y	TGAA	i	GATG	y	GTCG
*	AATG	:	CTAG	J	TTCC	Z	TAAG	j	GATT	z	GTGA
+	AAGA	;	CCGC	K	TACT	[	TCAT	k	GGGC	{	GGCT
,	AGAG	<	CACA	L	TATG	\\	TAAG	l	GTTG	|	GGTG
-	AAGC	=	CATA	M	TAGT	]	TCCA	m	GTGA	}	GAAC
.	ACAC	>	CTAC	N	TGTC	^	TGTT	n	GACT	~	GATG
/	ACGT	?	CCAG	O	TATT	_	TCCG	o	GCCG	\x7f	GAGT

Fig. 4.2 ASCII symbols and their corresponding DNA sequence

- Convert the plaintext message into an ASCII form and then convert it to 8 bits binary coded form (Fig. 4.2).
- Represent the binary data in the DNA coded form (A-00, C-01, G-10, T-11): *convert encode binary information into DNA strands.*
- Apply complementary rule to the sequence (A → C, C → G, G → T, T → A).
- Convert it back to binary.
- Generate random key and convert it into DND strands then into binary format. The random key has to be a number between 1 and 256. This random key determines the permutation of the four characters A, T, G, and C. For example, when the random key is 1, there is a table for the conversion of ASCII code to nucleotide sequences, and when it is 2, there is another table and so on.
- XORing the key with the data.

DNA decryption is the reverse operation of DNA encryption. The total key space results is approximating to 10^{23}. With such huge key space, the reliability and effectiveness of the algorithm are established as the *key* associated is quite unpredictable and resistant against brute force attacks. Figure 4.3 shows the results of encryption of an image using DNA cryptography.

4.2 Quantum Cryptography

Quantum cryptography uses physics to develop a cryptosystem completely secure against being compromised without the knowledge of the sender or the receiver of the messages. The word *quantum* itself refers to the most fundamental behavior of the smallest particles of matter and energy. In 1982, Richard Feynman came up with the idea of quantum computer, a computer that uses the effects of quantum mechanics to its advantage. In quantum cryptography, two remote parties can communicate

(a)   (b)

Fig. 4.3 (**a**) unencrypted image, (**b**) DNA-based encrypted image

securely by using the laws of quantum physics. Quantum cryptography is different
from traditional cryptographic systems in that it relies more on physics, rather than
mathematics, as a key aspectof its security model.

4.2.1 Properties of Quantum Information

Quantum cryptography rests on two pillars quantum mechanics [28–31]:

- *The Heisenberg uncertainty principle*: it is not possible to measure the quantum
  state of any system without disturbing that system. Thus, the polarization of a
  photon or light particle can only be known at the point when it is measured.
- *The photon polarization principle:* how light photons can be oriented or polarized in specific directions. Moreover, a photon filter with the correct polarization
  can only detect a polarized photon or else the photon will be destroyed.

Quantum computer exploits a kind of massive parallelism that cannot be
approached by any classical computer. So, it is faster [17]. Quantum technology is
a promising solution to overcome information security risks. Key distribution is one
of the most important challenges of cryptography. Quantum cryptography can help
in solving this problem. In quantum cryptography information is transmitted by
quantum bit, also called qubit, which is actually a single photon particle.

The current methods for breaking RSA are not very effective. One method is to
factor the N described by the public key. However, with the magnitude of the primes
chosen, factoring takes near-infinite time with current methods and technologies
(factoring time grows exponentially with input length in bits). In the present day,
RSA cannot be broken. However, theoretically it is vulnerable, if a fast algorithm of
semi-prime factoring was discovered. So, quantum computing is a threat for RSA
encryption.

For years, quantum computers have just been research, theory, and proposals. D-Wave is one of the companies that are making quantum computers a reality.

A regular bit is a transistor that registers either a high or low voltage, which corresponds to 1 or 0, respectively. A quantum bit is a 2-state quantum. Many things can be used as qubits, such as a photon's horizontal and vertical polarization or the spin up or spin down of an electron. Qubits also have very important properties [38]:

- *Superposition*: this is where a qubit is, while left unobserved, all of its possible states. Once observed, it will collapse into one of the possible states.
- *Entanglement*: This is where one qubit's state is linked to another. When entangled with each other, a change in one of the entangled qubits will change the other instantly.

4.2.2   Quantum Algorithms

The promise of quantum computing is that it will help us solve some of the world's most complex challenges. Quantum systems will have capabilities that exceed our most powerful supercomputers. The two basic algorithms of quantum cryptography are Shor's algorithm and the Grover's algorithm.

4.2.2.1   Grover's Algorithm

Grover's algorithm is a quantum algorithm that finds with high probability the unique input to a black box function that produces a particular output value.

4.2.2.2   Shor's Algorithm

Shor's algorithm provides a dramatic improvement in the efficiency of factoring large numbers. Thus, Shor's algorithm can be used to attack RSA encryption and related problems. It solves the following problem: Given an integer N, find its prime factors [39].

4.2.2.3   Quantum Cryptography Algorithm: The BB84 Protocol

The fundamental concept of quantum cryptography is sending secret key in the form of photons through an insecure channel. Binary data (zero and one) is encoded to a quantum state based on physics theories. Quantum cryptography is also well-known as quantum key distribution (*QKD*). There are two channels in this system. The first channel is used to transmit the quantum secret key with a single photon. The second channel is a public channel like a telephone line or the Internet used to exchange cryptography protocols. The lasers, specifically diode lasers, in the area of QKD [47, 48].

The BB84 protocol is the historical first protocol for quantum key, whose security is based on the principles of quantum mechanics, making it absolutely safe if there is no noise in the quantum channel. The absence of noise in a given situation assumes that the quantum state of particles does not change along the quantum channel. The BB84 protocol is formulated in the language of individual photons, although it can be applied to other realizations of a qubit.

The essence of the BB84 protocol is that one of the users (Alice) randomly selects a series of bits and a series of bases and then sends a user (Bob) a string of photons each of which encodes one bit from the selected string in the base corresponding to the prime number of that bit. In obtaining a photon, Bob randomly selects the measurement base for each photon and, independently of Alice, analogously interprets the result of his measurement for each photon in two ways, as a zero or one. In accordance with the laws of quantum mechanics and following the measuring of the diagonal photon in a rectangular base, its polarization turns into the horizontal or vertical line and vice versa, with random results. In this way, Bob obtains the results coinciding with the state of the photons sent in about half the cases (50%), that is, when he correctly hits the base.

The next stage of the protocol is realized via a public channel, through which Alice and Bob can openly convey classical information to each other. At this stage, we assume that Eva can listen to the announcements by both parties, but she cannot change them or send notifications instead of them. To begin with, Alice and Bob determine (via a public channel) which photons were successfully obtained by Bob and which of them were measured in the correct base. After that, Alice and Bob have the same bit values encoded in these photons, regardless of the fact that this information has never been established in the open communication channel. In other words, each of these photons carries a bit of random information, which is known only to Alice and Bob and no one else. Information about the photons measured in the wrong base is rejected, so Alice and Bob get the so-called sieved key, which, in the event that Eva did not intercept the information, should be the same for both parties. Suppose Eva is eavesdropping on a quantum channel. Due to the random selection of a rectangular or diagonal base, Eva influences the information in such a way that it changes the bits of the sieved key, which would have to be the same for Alice and Bob if there was no Eve [49].

4.2.3 Quantum Cryptography Challenges

When we compare post-quantum cryptography with the currently used asymmetric algorithms, we find that post-quantum cryptography mostly have larger key and signature sizes and require more operations and memory. Still, they are very practical for everything except perhaps very constrained Internet of Things devices and radio. Some other challenges are summarized below:

- Expensive: need specialized hardware.
- Complex.
- Difficult to implement over long distance.
- Subjected to decoherence. Qubits can retain their quantum state for a short period of time.
- Quantum algorithms are mainly probabilistic. This means that in one operation, a quantum computer returns many solutions where only one is the correct.

4.3 Chaotic Cryptography

Embedded systems are the driving force for most technological development in many domains such as automotive and healthcare. An example of embedded system architecture is shown in Fig. 4.4.

Almost 10% of all embedded system products are counterfeit which leads to huge revenue loss. Several attacks methods have been developed, which made it possible to learn the ROM-based keys. Protecting soft IPs is much more challenging because it can be easily copied and even be sold at lower levels of abstraction [11]. Types of attacks on the hardware/software level can be classified into virus/worms, reverse engineering, fault injection, memory modification (Trojan insertion, bus modification, side channel, and bus probing).

Many strong ciphers have been applied widely, such as DES, AES, and RSA. But most of them cannot be directly used to encrypt real-time embedded systems because their encryption speed is not fast enough and they are computationally intensive. So, in this work we present a fast chaotic-based encryption algorithm which is suitable for real-time embedded systems in terms of performance, area, and power efficiency.

4.3.1 Chaotic Theory

All systems can be basically divided into three types [42]:

1. Deterministic systems: these are systems for which for a given set of conditions the result can be predicted and the output does not vary much with change in initial conditions. Examples are computers.
2. Stochastic/random systems: these systems, which are not as reliable as deterministic systems. Their output can be predicted only for a certain range of values. Examples are genetic algorithms.
3. Chaotic systems: these systems are the most unpredictable of the three systems. Moreover they are very sensitive to initial conditions, and a small change in initial conditions can bring about a great change in its output. Examples of chaotic systems are the solar system, population growth, stock market, and weather.

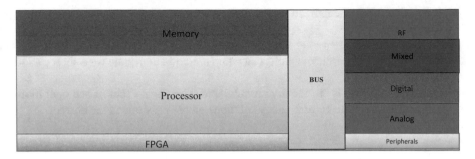

Fig. 4.4 Example of embedded system architecture

Chaos is derived from the Greek word "Χαos," which is meaning a state without predictability or order. A chaotic system is a nonlinear, dynamical, and deterministic system which has high sensitive to initial conditions of the system. Chaos system is deterministic system with small change in input results in enormous change in the output, so the system looks as if it is random and prediction becomes impossible (it looks like a noise). It is like butterfly effect. Due to these properties, chaos theory has been used in cryptography/encryption. In this work, chaotic theory is used for providing security at HW level.

4.3.2   Chaotic Encryption System

The proposed chaotic-based encryption is taken from chaotic interleaving in communication [12–14]. Here, we proposed to use the algorithm for IP protection in embedded systems. Chaotic encryption of an $N \times N$ square matrix of data can be summarized as follows

1. An $N \times N$ square matrix is divided into k vertical rectangles of height N and width n_i such that $n_1 + n_2 + \ldots + n_k = N$.
2. These vertical rectangles are stretched in the horizontal direction and contracted vertically to obtain $n_i \times N$ horizontal rectangle.
3. These rectangles are stacked such as the left one is put at the bottom and the right one at the top.
4. Each vertical rectangle $n_i \times N$ is divided into n_i boxes of dimensions $\dfrac{N}{n_i} \times n_i$ containing exactly N points.
5. Each of these boxes is mapped column by column into a row of data items. Inside each rectangle, the scan begins from the bottom left corner toward upper elements.

Figure 4.5 shows an example of chaotic encryption of an (8×8) square matrix. The secret key $S_{\text{key}}(n_1, n_2, n_3) = (2, 4, 2)$. The security abstraction levels are shown in Table 4.1. The proposed algorithm is working at the algorithm level.

b1	b2	b3	b4	b5	b6	b7	b8
b9	b10	b11	b12	b13	b14	b15	b16
b17	b18	b19	b20	b21	b22	b23	b24
b25	b26	b27	b28	b29	b30	b31	b32
b33	b34	b35	b36	b37	b38	b39	b40
b41	b42	b43	b44	b45	b46	b47	b48
b49	b50	b51	b52	b53	b54	b55	b56
b57	b58	b59	b60	b61	b62	b63	b64

(a)

b1	b2	b3	b4	b5	b6	b7	b8
b9	b10	b11	b12	b13	b14	b15	b16
b17	b18	b19	b20	b21	b22	b23	b24
b25	b26	b27	b28	b29	b30	b31	b32
b33	b34	b35	b36	b37	b38	b39	b40
b41	b42	b43	b44	b45	b46	b47	b48
b49	b50	b51	b52	b53	b54	b55	b56
b57	b58	b59	b60	b61	b62	b63	b64

(b)

b31	b23	b15	b7	b32	b24	b16	b8
b63	b55	b47	b39	b64	b56	b48	b40
b11	b3	b12	b4	b13	b5	b14	b6
b27	b19	b28	b20	b29	b21	b30	b22
b43	b35	b44	b36	b45	b37	b46	b38
b59	b51	b60	b52	b61	b53	b62	b54
b25	b17	b9	b1	b26	b18	b19	b2
b57	b49	b41	b33	b58	b50	b42	b34

(c)

Fig. 4.5 An illustrating example of the proposed chaotic map. (**a**) Raw data 8 bits × 8bits, (**b**) the chaotic map creation 2 × 4 × 2, (**c**) the encrypted matrix

Table 4.1 The security abstraction levels

Security abstraction level	Security objective	Side channel attack
Protocol	Authenticated communications	Main-in-the-middle, traffic analysis
Algorithm	Encryption/hashing	Known-plaintext, known-cryptext
Architecture	Functional integration	Stack smashing
Micro-architecture	Architecture integration	Bus probing
Circuit	Implementation	Differential power analysis

To evaluate the resistance against security threats, security analysis of proposed encryption algorithm is done. The system proves to be efficient against different types of attacks. If the matrix size becomes 1024 × 8, then the total key space comes out to be sufficient to resist brute force attacks. Based on the proposed architecture, we can represent the key space as a series:

$$\text{Key}_{space} = (n-1).\left(1 + \sum_{j=1}^{j=n-2i=2^{(n-j-1)}-1} \sum_{i=1} \left(2^{n-j} - 2.j\right)\right) \tag{4.1}$$

Assuming the key size = 1024 = 2^{10} ($n = 10$), so: key$_{space}$ ≅ 1.575 M.

4.3.3 Hardware Implementation of Chaotic Algorithm

First, the proposed algorithm is implemented in MATLAB for performance evaluation against conventional encryption methods. Then it is implemented using Verilog. The key can be hardwired. Chaotic encryption is a symmetric encryption scheme, since it uses same parameters/MAP for encryption and decryption process. Block diagram of the proposed encryption algorithm is shown in Fig. 4.6. The proposed encryption architecture consists of four sub-modules:

- FIFO: to store the incoming data.
- Vector to matrix: as Verilog does not accept 2D matrix as an input or output port (Fig. 4.7).
- Reshape: applying the chaotic map.
- Matrix to vector: convert matrix to vector.

The decryption process is the inverse operation. The architecture of the decryption process is shown in Fig. 4.8.

It consists of the following:

- FIFO: to store the incoming data.
- Vector to Matrix: as Verilog does not accept 2D matrix as an input or output port.
- Reshape: applying the chaotic map.
- Matrix to vector: convert matrix to vector.

Fig. 4.6 Block diagram of the proposed encryption algorithm

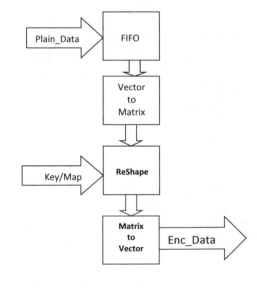

Fig. 4.7 Vector to matrix block diagram as Verilog does not accept 2D matrix as an input or output port

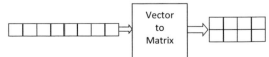

Fig. 4.8 Block diagram of
the proposed decryption
algorithm

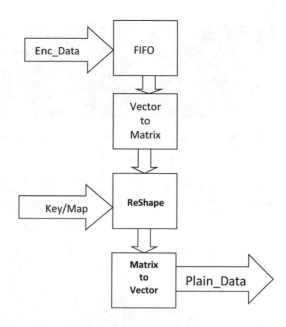

Fig. 4.9 FPGA evaluation board used to perform the proposed algorithm implementation

4.3.4   Evaluation of the Proposed Algorithm

We synthesized the design over a Xilinx Virtex-6 FPGA as depicted in Fig. 4.9. A
direct implementation of the design gave a clock frequency of 400 MHz. The
throughput is 3.2 Gbps. Compared to conventional encryption methods, the performance of our proposed method is better. A snapshot of the simulation results is

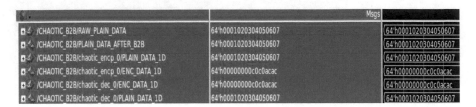

Fig. 4.10 Simulation results for a back-to-back configuration of the encryption/decryption modules

Table 4.2 Comparison between our proposed algorithm and AES

File (KB)	Decryption time (m Sec)	
	Our proposed chaotic encryption	AES
1	1.5	3
2	2.8	6
3	4.7	9
4	6	12
5	7.7	15
6	9.1	18

Table 4.3 Area, delay, and power overhead for our proposed encryption method

Our proposed chaotic encryption			AES		
Area	Delay (ns)	Power (nW)	Area	Delay (ns)	Power (nW)
5%	0.1	0.3	40%	0.4	2.6

Table 4.4 Area, delay, and power overhead for our proposed encryption method

Platform	Virtex 7
LUTs	600
Registers	54
Total DSPs	–
Freq (MHz)	400
Throughput (Gbps)	3.2
Slices	180

shown in Fig. 4.10 where a back-to-back configuration of the encryption/decryption modules is used to ensure that the decrypted data is the same as the plain data. There is a trade-off between the area/latency/power overhead and the provided level of security. Our proposed algorithm provides a good level of security with low area/power/latency overhead. The area is about 5% of AES area. The power consumption is less than AES. The comparison between our proposed algorithm and AES in terms of decryption time for different file sizes is shown in Table 4.2, where our proposed algorithm shows better and fast performance. Moreover, area, delay, and power overhead for our proposed encryption method are shown in Tables 4.3 and 4.4. The resources used by the encryptor including the total number of lookup tables (LUTs), slice registers, and digital signal processing blocks (DSPs), as well as the throughput obtained, are shown in Table 4.5.

Table 4.5 Comparison with related work

Metric	[1]	[2]	[3]	This work
Latency (clock cycles)	33	133	295	21
Max Freq (MHz)	306	375	236	400
Throughput (Mbps)	595	180	50	600
FPGA slices (Vertex 7)	112	124	62	40
Power (mW)	–	245	–	210

4.4 Lightweight Cryptography

Lightweight cryptography works between the trade-offs of security, cost, and performance and is focused at devices and systems on edge. The increase in Internet-connected devices requires to build smarter systems that are secure using low-cost hardware solutions. The symmetric and asymmetric ciphers are essentially a major topic of study in hardware cryptography, each having a different set of applications. Hardware for asymmetric ciphers are more complex than symmetric ones and consume more area on chip and power. For example, in terms of computational complexity, symmetric cipher such as AES algorithm is about 1000 much faster than an optimized elliptic curve cryptography that is an asymmetric algorithm. The Internet of Things (IoT) is one of the most promising research topics in the engineering field. IoT is believed to facilitate the way people live in the near future by distantly connecting objects with each other and establishing communication channels between them. According to Cisco's Internet of Things Group (IoTG), the number of connected devices is expected to reach 50 billion by 2020. IoT has much potential to revolutionize the industry and everyday life in the near future, but some challenges hinder its advancements such as power consumption and security issues. Without sufficient security and privacy, all the benefits of IoT could prove disadvantageous if misused. Different algorithms have been presented in the literature that meets security requirements. Most of the studies have focused on popular algorithms such as AES, Rijndael, DES, Twofish, RSA, and more. However, IoT low area and power requirements make these algorithms unsuitable [22]. So, more IoT-oriented algorithms have been presented to provide better performance in terms of power and area; those are known as lightweight cryptographic algorithms such as PRESENT, RECTANGLE, SIT, HIGHT, CLEFIA, SPECK, SIMON, and KHUDRA algorithm [1–10].

4.4.1 PRESENT Algorithm

PRESENT is an ultralow power encryption algorithm that is based on substitution permutation (SP) network. PRESENT has a 64 bit input plaintext and either 80 bit or 128 bit key. PRESENT was standardized by NIST in 2012, which provide the algorithm more credibility in its use [46]. PRESENT is a 31-round operation in

i	0	1	2	3	4	5	6	7	8	9	10	11	12	13	14	15
$P(i)$	0	16	32	48	1	17	33	49	2	18	34	50	3	19	35	51
i	16	17	18	19	20	21	22	23	24	25	26	27	28	29	30	31
$P(i)$	4	20	36	52	5	21	37	53	6	22	38	54	7	23	39	55
i	32	33	34	35	36	37	38	39	40	41	42	43	44	45	46	47
$P(i)$	8	24	40	56	9	25	41	57	10	26	42	58	11	27	43	59
i	48	49	50	51	52	53	54	55	56	57	58	59	60	61	62	63
$P(i)$	12	28	44	60	13	29	45	61	14	30	46	62	15	31	47	63

Fig. 4.11 Permutation

x	0	1	2	3	4	5	6	7	8	9	A	B	C	D	E	F
$S[x]$	C	5	6	B	9	0	A	D	3	E	F	8	4	7	1	2

Fig. 4.12 Substitution

Algorithm 3 PRESENT encryption algorithm

Require: $Plaintext[64]$, $Key[80]$
Ensure: $Ciphertext[64]$
  $Round \leftarrow 1$
  $State \leftarrow Plaintext$
  while $R \neq 31$ **do**
    $RoundKey_R \leftarrow PRESENTgenerateKey_{80}(Key, R)$
    $State \leftarrow State \oplus RoundKey_R$
    $State \leftarrow SBoxLayer(State)$
    $State \leftarrow PermutationLayer(State)$
    $R \leftarrow R + 1$
  end while
  $lastRoundKey \leftarrow generateKey(Key, R)$
  $State \leftarrow State \oplus lastRoundKey$
  $Ciphertext \leftarrow State$

Fig. 4.13 Pseudo-code for PRESENT algorithm

which an XOR operation is introduced with round Key Ki. It consists of linear transformation called permutation (Fig. 4.11) and nonlinear transformation called substitution (Fig. 4.12). The substitution and permutation are performed once every round. A new key is generated for each round. Decryption is the inverse of the encryption process [18]. Pseudo-code for PRESENT algorithm is shown in Fig. 4.13. To test the proposed encryption/decryption algorithm, we connect them back to back (Fig. 4.14). The power consumption is about 210 mW which is less power than the related work. Moreover, the overall area is less than the related work. For encryption (Fig. 4.15), there are four main blocks for the PRESENT: AddRoundKey, S-box, P-layer, and key schedule. They are working in the following manner [51, 52].

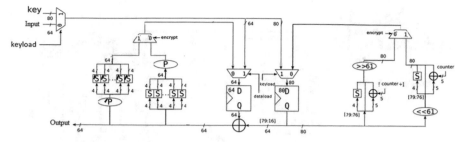

Fig. 4.14 Back-to-back configuration

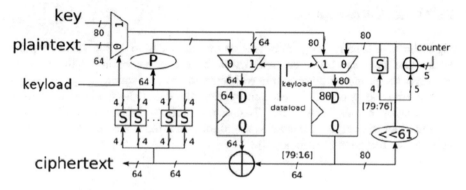

Fig. 4.15 PRESENT encryption core

- Step1: plaintext and the key are stored in a register.
- Step2: the plaintext is XORed with the key.
- Step3: a substitutional step is done to provide the confusion needed.
- Step4: the data is permuted and stored in the register and the counter is increased by 1 and so on.

For the decryption core (Fig. 4.16), the overall steps are similar to the ones explained in the case of encryption but with some key differences.

- The s-box, permutation, and key scheduling units are replaced with their inverse modules.
- The inverse permutation layer is carried out before the inverse s-box layer.

4.4.2   SIT Algorithm

SIT is a symmetric key block cipher that constitutes of 64 bit key and plaintext. In symmetric key algorithm, the encryption process consists of encryption rounds; each round is based on some mathematical functions to create confusion and diffusion. Increase in number of rounds ensures better security but eventually results in increase in the consumption of constrained energy [21]. The cryptographic

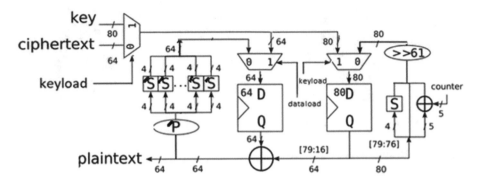

Fig. 4.16 PRESENT decryption core

algorithms are usually designed to take on an average 10 to 20 rounds to keep the encryption process strong enough that suits the requirement of the system. However the proposed algorithm is restricted to just five rounds only; to further improve the energy efficiency, each encryption round includes mathematical operations that operate on 4 bits of data. To create sufficient confusion and diffusion of data in order to confront the attacks, the algorithm utilizes the Feistel network of substitution diffusion functions [15].

4.4.3   HIGHT Algorithm

High security lightweight (HIGHT) algorithm is based on Feistel network structure instead of SPN. HIGHT operates on a 64 bit block size with 128 bit key size. The algorithm is comprised of 32 rounds, each is based on basic operations such as XOR and addition mod 28 [19].

4.4.4   KHUDRA Algorithm

KHUDRA is an FPGA-oriented lightweight algorithm. It's optimized for balancing LUTs and registers to minimize the FPGA slices. The algorithm is based on recursive Feistel [20].

4.4.5   CAMELLIA Algorithm

It is somehow similar to the standard AES, as it's a symmetric key block cipher with a fixed block size of 128 bits and three different key sizes of 128, 192, and 256 bits. Unlike other algorithms that focus on hardware implementation,

CAMELLIA was designed for both software and hardware. It can be used for both low-cost and high-speed applications [40].

4.4.6   Attribute-Based Encryption (ABE)

ABE is a type of encryption that provides the IoT network with privacy and security through a policy between the attributes of the users in the system. ABE consists of two types, key policy ABE (KP-ABE) and ciphertext policy ABE (CP-ABE). In CP-ABE, the sender's data access policy is embedded in the ciphertext, and a recipient's attributes are associated with its private keys. A sender can decrypt the ciphertext only if the attributes associated with its private key satisfy the access policy embedded in the encrypted data [44, 45].

4.5   Blockchain Cryptography

Blockchain is a distributed database that allows direct transactions between two parties without the need for an authoritative mediator. Blockchain is a way to encapsulate transactions in the form of blocks where blocks are linked through the cryptographic hash, hence forming a chain of blocks (Fig. 4.17). Blockchain is used for integrity as depicted in Fig. 4.18. Blockchain relies on different constituents which serve different purposes. The blockchain consists of a sequence of blocks that are stored on and copied between publicly accessible servers. Each block consists of four fundamental elements: the hash of the preceding block; the data content of the block (i.e., the ledger entries); the nonce that is used to give a particular form to the hash; and the hash of the block. By including the hash of the preceding block, each successive block strengthens the authenticity claim for the preceding block.

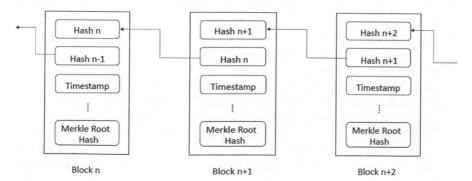

Fig. 4.17 Basic blockchain structure. A block has a hash (fingerprint) which is unique to each block. It identifies a block and all of its contents, and it's always unique. So once a block is created, any change inside the block will cause the hash to change

Fig. 4.18 Blockchain: integrity

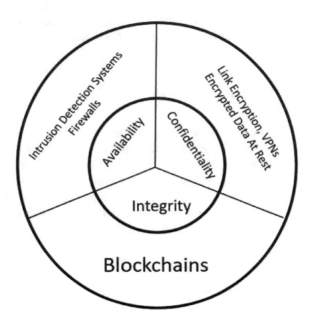

Blocks early in the chain cannot be modified without modifying all subsequent blocks, or the modification will appear as an inconsistency in the hashes. Similarly, adding the data to the hash makes the data unmodifiable without breaking the consistency of the block sequence. Adding a nonce that is used to impose a signature structure to the hash requires significant work to be performed to generate a new block [41]. Assume an attacker is able to change the data present in the block n. Correspondingly, the hash of the block also changes. But, block $n + 1$ still contains the old hash of the block n. This makes block $n + 1$ and all succeeding blocks invalid as they do not have correct hash the previous block. Blockchain technology has become so popular due to the following advantages [43]:

- *Resilience:* Blockchain is often replicated architecture. The chain is still operated by most nodes in the event of a massive attack against the system.
- *Time reduction:* In the financial industry, blockchain can play a vital role by allowing the quicker settlement of trades as it does not need a lengthy process of verification, settlement, and clearance because a single version of agreed-upon data of the share ledger is available between all stack holders.
- *Reliability:* Blockchain certifies and verifies the identities of the interested parties. This removes double records, reduces rates, and accelerates transactions.
- *Unchangeable transactions:* By registering transactions in chronological order, blockchain certifies the unalterability of all operations which means when any new block has been added to the chain of ledgers, it cannot be removed or modified.
- *Fraud prevention:* The concepts of shared information and consensus prevent possible losses due to fraud or embezzlement. In logistics-based industries, blockchain as a monitoring mechanism acts to reduce costs.

- *Security:* Attacking a traditional database is the bringing down of a specific target. With the help of distributed ledger technology, each party holds a copy of the original chain, so the system remains operative, even the large number of other nodes fall.
- *Transparency:* Changes to public blockchain are publicly viewable to everyone. This offers greater transparency, and all transactions are immutable.
- *Collaboration:* Allows parties to transact directly with each other without the need for mediating third parties.
- *Decentralized:* There are standard rules on how every node exchanges the blockchain information. This method ensures that all transactions are validated and all valid transactions are added one by one [50].

4.5.1   Limitations of Blockchain Technology Can Be Summarized as Follows

- *Higher costs:* Nodes seek higher rewards for completing transactions in a business which work on the principle of supply and demand.
- *Slower transactions:* Nodes prioritize transactions with higher rewards; backlogs of transactions build up.
- *Smaller ledger:* It is not possible to a full copy of the blockchain, potentially which can affect immutability, consensus, etc.
- *Transaction costs and network speed:* The transactions cost of Bitcoin is quite high after being touted as "nearly free" for the first few years.
- *Risk of error:* There is always a risk of error, as long as the human factor is involved. In case a blockchain serves as a database, all the incoming data has to be of high quality. However, human involvement can quickly resolve the error.
- *Wasteful:* Every node that runs the blockchain has to maintain consensus across the blockchain. This offers very low downtime and makes data stored on the blockchain forever unchangeable. However, all this is wasteful, because each node repeats a task to reach consensus.

4.5.2   Prime Number Factorization

The mathematical principle behind prime number factorization is that any number, no matter how large, can be produced by multiplying prime numbers. It's relatively easy to produce any number using prime numbers. However, it's vastly more difficult to reverse the process and work out which prime numbers were multiplied to produce a particular value once the numbers become large. This reversal is called prime number factorization (Fig. 4.19). *Blockchain cryptography relies on prime number* factorization for linking the public and private key. The prime number factors of the public key are what form the private key [16].

Fig. 4.19 Prime number
factors example

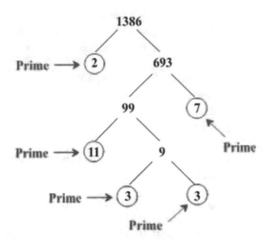

4.5.3   Applications of Blockchain Cryptography

4.5.3.1   Money Transfer

Blockchain is a groundbreaking technology that optimizes the way money is trans-
ferred and transactions are processed. While it has been used in many fields since its
introduction in 2009, blockchain technology is still most widely used in money
transfers and transaction reconciliation.

4.5.3.2   Smart Contract

The new key concepts are smart contracts, small computer programs that "live" in
the blockchain. They are free computer programs that execute automatically and
check conditions defined earlier like facilitation, verification, or enforcement. It is
used as a replacement for traditional contracts.

4.5.3.3   Safety of Food

Food companies implement traceability because they see that the consumers require
transparency and credibility. Blockchain's immutability helps them to prove that the
information the different supply chain companies provide is uncorrupted.

4.5.3.4   Cryptocurrency

Bitcoin, the first decentralized cryptocurrency, has gained a large attention since its
inception in 2009. Built upon blockchain technology, it has established itself as the
leader of cryptocurrencies and shows no signs of slowing down. Instead of being

based on traditional trust, the currency is based on cryptographic proof which provides many advantages over traditional payment methods (such as Visa and Mastercard) including high liquidity and lower transaction costs [32]. The blockchain is the technology behind Bitcoin. Bitcoin is the digital token, and blockchain is the ledger that keeps track of who owns the digital tokens. You can't have Bitcoin without blockchain, but you can have blockchain without Bitcoin.

4.6 Conclusions

This chapter discusses the cutting-edge cryptographic techniques such as quantum cryptography, DNA cryptography, chaotic cryptography, lightweight cryptography, and blockchain cryptography. All these cryptography techniques are promising and rapid emerging fields in data security.

References

1. A. Bogdanov et al., PRESENT: An ultra-lightweight block cipher, in *Cryptographic Hardware and Embedded Systems – CHES*, (2007)
2. J. Borghoff, L.R. Knudsen, G. Leander, S.S. Thomsen, Cryptanalysis of PRESENT-like ciphers with secret S-boxes, in *International Workshop on Fast Software Encryption*, (Springer, Berlin, Heidelberg, 2011), pp. 270–289
3. J. Pospiil, M. Novotný, Evaluating cryptanalytical strength of lightweight cipher present on reconfigurable hardware, in *Digital System Design (DSD), 2012 15th Euromicro Conference*, (IEEE, 2012), pp. 560–567
4. S.S. Rekha, P. Saravanan, Low cost circuit level implementation of PRESENT-80 S-BOX, in *International Symposium on VLSI Design and Test*, (Springer, Singapore, 2017), pp. 354–362
5. C.A. Lara-Nino, M. Morales-Sandoval, A. Diaz-Perez, Novel FPGA-based low-cost hardware architecture for the PRESENT block cipher, in *2016 Euromicro Conference on Digital System Design (DSD)*, (IEEE, 2016), pp. 646–650
6. D. Bellizia, G. Scotti, A. Trifiletti, Implementation of the present-80 block cipher and analysis of its vulnerability to side channel attacks exploiting static power, in *Mixed Design of Integrated Circuits and Systems, 2016 MIXDES-23rd International Conference*, (IEEE, 2016), pp. 211–216
7. C. Andrés, M.S. Miguel, D.P. Arturo, An evaluation of AES and present ciphers for lightweight cryptography on smartphones, in *Electronics, Communications and Computers (CONIELECOMP), 2016 International Conference*, (IEEE, 2016), pp. 87–93
8. J.G. Pandey, T. Goel, A. Karmakar, An efficient VLSI architecture for PRESENT block cipher and its FPGA implementation, in *International Symposium on VLSI Design and Test*, (Springer, Singapore, 2017), pp. 270–278
9. C.A. Lara-Nino, A. Diaz-Perez, M. Morales-Sandoval, Lightweight hardware architectures for the PRESENT cipher in FPGA. IEEE Trans. Circ. Syst. Reg. Papers **64**(9), 2544–2555 (2017)
10. J.J. Tay, M.L.D. Wong, M.M. Wong, C. Zhang, I. Hijazin, Compact FPGA implementation of PRESENT with Boolean S-Box, in *Quality Electronic Design (ASQED), 2015 6th Asia Symposium*, (IEEE, 2015), pp. 144–148
11. K. Salah, *IP Cores Design from Specifications to Production: Modeling, Verification, Optimization, and Protection* (Springer, 2016)

12. A.Z. El Hamid, A. El-Henawy, H. El-Shenawy, Performance evaluation of chaotic interleaving with FFT and DWT OFDM, in *29th National Radio Science Conference (NRSC)*, (IEEE, 2012)
13. A.M. El-Bendary, A. Abou El-Azm, An efficient chaotic interleaver for image transmission over IEEE 802.15. 4 Zigbee network. J. Telecommun. Inform. Tech. (2011)
14. E.S. Hassan, S.E. El-Khamy, M. Dessouky, A chaotic interleaving scheme for the continuous phase modulation based single-carrier frequency-domain equalization system. Wirel. Pers. Commun. **62**(1), 183–199, Springer, (2012)
15. M. Usman, I. Ahmedy, M. Imran Aslamy, S. Khan, U.A. Shahy, SIT: A lightweight encryption algorithm for secure internet of things. Int. J. Adv. Comput. Sci. Appl. **8**(1) (2017)
16. https://coincentral.com/blockchain-cryptography-quantum-machines/
17. J.B. Altepeter, *A Tale of Two Qubits: How Quantum Computers Work*. Ars Technica (Online Magazine), January 18, 2010
18. A. Bogdanov et al., PRESENT: An ultra-lightweight block cipher, in *Cryptographic Hardware and Embedded Systems – CHES 2007 Lecture Notes in Computer Science*, (Springer, 2007), pp. 450–466
19. D. Hong et al., HIGHT: A new block cipher suitable for low resource device, in *Cryptographic Hardware and Embedded Systems – CHES 2006 Lecture Notes in Computer Science*, (2006), pp. 46–59
20. S. Kolay, D. Mukhopadhyay, Khudra: A new lightweight block cipher for FPGAs, in *SPACE, Vol 8804 of LNCS*, (Springer, 2014), pp. 126–145
21. M. Usman, I. Ahmed, M. Imran, S. Khan, U. Ali, SIT: A lightweight encryption algorithm for secure internet of things. Int. J. Adv. Comput. Sci. Appl. **8**(1) (2017)
22. S. Koteshwara, A. Das, Comparative study of authenticated encryption targeting lightweight IoT applications. IEEE Design Test **34**(4), 26–33 (2017)
23. B.B. Raj, J. Frank, T. Mahalakshmi, Secure data transfer through DNA cryptography using symmetric algorithm. Int. J. Comput. Appl. **133**(2), 0975–8887 (2016)
24. A. Roy, A. Nath, DNA encryption algorithms: Scope and challenges in symmetric key cryptography. Int. J. Innov. Res. Adv. Eng., ISSN: 2349-2763, **3**(11) (2016)
25. W. Stallings, *Cryptography and Network Security* (3rd, Prentice Hall International, 2003)
26. N.S. Kolte, K.V. Kulhalli, S.C. Shinde, DNA cryptography using index-based symmetric DNA encryption algorithm. Int. J. Eng. Res. Tech., ISSN 0974-3154, **10**(1) (2017)
27. M. Najaftorkaman, N.S. Kazazi, A method to encrypt information with DNA-based cryptography. Int. J. Cyber Sec. Digital Foren., The Society of Digital Information and Wireless Communications, (2015)
28. Z. Kirsch, *Quantum Computing: The Risk to Existing Encryption Methods*, Ph.D. dissertation, Tufts University, Massachusetts, 2015
29. L.S. Bishop, S. Bravyi, A. Cross, J.M. Gambetta, J. Smolin, Quantum volume, Technical report, 2017, Tech. Rep., 2017
30. D. Bernstein, E. Dahmen, Buch, Introduction to post-quantum, in *Cryptography*, (Springer-Verlag, Berlin Heidelberg, 2010)
31. W. Buchanan, A. Woodward, Will quantum computers be the end of public key encryption? J. Cyber Sec. Tech. **1**(1), 1–22 (2016)
32. A. Hencic, C. Gourieroux, Noncausal autoregressive model in application to Bitcoin/USD exchange rate, in *Econometrics of Risk*, (Springer, Berlin, 2014), pp. 17–40
33. E.S. Babu, M.H.M.K. Prasad, C.N. Raju, Inspired pseudo biotic DNA based cryptographic mechanism against adaptive cryptographic attacks. Int. J. Network Sec. **18**(2), 291–303 (2016)
34. M.S.S. Basha, I.A. Emerson, R. Kannadasan, Survey on molecular cryptographic network DNA (MCND) using big data, in *Procedia Computer Science of 2nd International Symposium on Big Data and Cloud Computing (ISBCC'15)*, vol. 50, (2015), pp. 3–9
35. M. Bhavithara, A.P. Bhrintha, A. Kamaraj, DNA-based encryption and decryption using FPGA. Int. J. Curr. Res. Mod. Edu. (IJCRME'16), 89–94 (2016)

36. N.S. Kazazi, M.R.N. Torkaman, A method to encrypt information with DNA-based cryptography. Int. J. Cyber Sec. Dig. Foren. (IJCSDF'15) **4**(3), 417–426 (2015)
37. T. Mahalaxmi, B.B. Raj, J.F. Vijay, Secure data transfer through DNA cryptography using a symmetric algorithm. Int. J. Comput. Appl. **133**(2), 19–23 (2016)
38. Y. Hashimoto, Multivariate public key cryptosystems, in *Mathematical Modelling for Next-Generation Cryptography*, (Springer, 2018), pp. 17–42
39. P.W. Shor, Algorithms for quantum computation: Discrete logarithms and factoring, in *Proceedings of the 35th Annual Symposium on Foundations of Computer Science (SFCS'94)*, (IEEE, 1994), pp. 124–134
40. A. Satoh, S. Morioka, Hardware-focused performance comparison for the standard block ciphers AES, camellia, and triple DES. Lect. Notes Comput. Sci. Inform. Sec., Springer,, 252–266 (2003)
41. K.S. Mohamed, *Neuromorphic Computing and Beyond: Parallel, Approximation, Near Memory, and Quantum* (Springer Nature, 2020)
42. K. Salah, Real time embedded system IPs protection using chaotic maps, in *IEEE 8th Annual Ubiquitous Computing, Electronics and Mobile Communication Conference (UEMCON)*, (IEEE, 2017), p. 2017
43. https://www.guru99.com/blockchain-tutorial.html
44. X. Wang, J. Zhang, E. Schooler, M. Ion, Performance evaluation of attribute-based E encrypt ion: Toward data privacy in the IoT. IEEE Int. Conf. Commun. (ICC), 725–730 (2014)
45. M. Yagisawa, Key distribution system and attribute-based encrypt ion on non-commutative ring. Cryptol. ePrint Arch., Report 2012/24, (2012)
46. A. Bogdanov, L.R. Knudsen, G. Leander, C. Paar, A. Poschmann, M.J.B. Robshaw, Y. Seurin, C. Vikkelsoe, PRESENT: An ultra-lightweight block cipher, in *Proceeding of Cryptographic Hardware and Embedded Systems—CHES 2007*, (Springer), pp. 450–466
47. A. Singh, Centralized key distribution on quantum cryptography. Int. J. Comput. Sci. Mob. Comput. (IJCSMC) **6** (2017)
48. Y. Wang, K. She, A practical quantum public key encryption model. Int. Conf. Inform. Manag. (2017)
49. Stevo Jacimovski "on quantum cryptography" 2019
50. H.Q. Wang, T. Wu, Cryptography in Blockchain. J. Nanjing Univ. Posts Telecommun. **37**, 61–67 (2017)
51. R.H. Weber, Internet of Things—New security and privacy challenges. Comp. Law Sec. Rev. **26**(1), 23–30 (2010)
52. A. Ukil, J. Sen, S. Koilakonda, Embedded security for Internet of Things, in *Proceedings of 2nd National Conference on Emerging Trends and Applications in Computer Science (NCETACS)*, (2011), pp. 1–6

Chapter 5
Data Hiding: Steganography and Watermarking

5.1 Introduction

In era of information society, protection system can be classified into more specific as hiding information (steganography or watermarking) or encryption information (cryptography) or a combination between them. Cryptography and steganography are well-known and broadly used techniques that use information in order to cipher or cover their existence, respectively. Comparison between cryptography and information hiding techniques is shown in Fig. 5.1. Information hiding is a technique of hiding secret using redundant cover data such as images, audios, movies, documents, etc.

5.2 Steganography

The word "steganography" comes from the Greek words "stegano" which means "covered" and "graphie" means "writing." Steganography is the art of hiding the existence of a message between sender and intended recipient. It hides secret messages in various types of files such as text, images, audio, and video. Steganography hides encrypted messages in such a way that no one would even suspect that an encrypted message even exists in the first place [2]. Because there are unused/redundant data bits in digital media, that changing them will be imperceptible. Now, there are several techniques to conceal data in the normal files. The easiest and simplest method to hide secret information in a digital media is the least significant bit (*LSB*) coding. This technique is done by replacing the least significant bit of each sample with a bit of the secret data. Steganography is hiding data within data. Steganography is often linked to cryptography; however the two are not mutually exclusive. One might use steganography in conjunction with encryption in order to

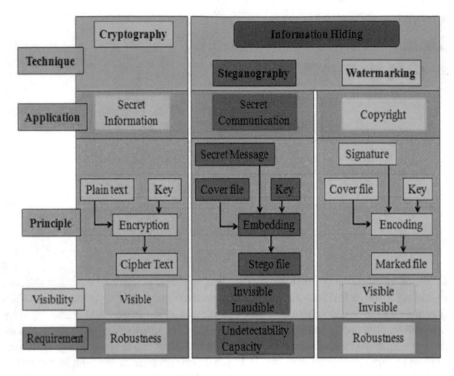

Fig. 5.1 Comparison between cryptography and information hiding techniques [1]

Table 5.1 Comparison between steganography and cryptography

Comparison	Steganography	Cryptography
Basic	It is known as cover writing	It means secret writing
Goal	Secret communication	Data protection
Structure of the message	Not altered	Altered only of the transmission
Popularity	Less popular	More commonly used
Relies on	Key	No parameters
Supported security principles	Confidentiality and authentication	Confidentiality, data integrity, authentication, and non-repudiation
Techniques	Spatial domain, transform domain, model-based, and ad hoc	Transposition, substitution, stream cipher, block ciphers
Implemented on	Audio, video, image, text	Only on text files
Types of attack	Steganalysis	Cryptanalysis

deliver a secret message to a recipient without drawing attention to the fact that a message was sent at all. Comparison between steganography and cryptography is shown in Table 5.1. Regarding the strength of steganography, firstly, steganography will make it difficult for attackers to realize that a message is being passed. Further, if he is somehow able to detect that a stego-image is being passed, then it would be difficult for him to recognize the real ciphertext in sense that which bits of pixels are used for encoding the message into the picture. Even after attacker is able to detect the ciphertext, it would be very difficult for him to decrypt it, due to the diffusion and confusion in the ciphertext [3, 4].

5.2.1   Steganography in Digital Media

5.2.1.1   Text Steganography

The techniques in text steganography are number of tabs, white spaces, and capital letters, just like Morse code is used to achieve information hiding [5, 6]. Various techniques are used to hide the data in the text such as format-based method, random and statistical generation, and linguistic method.

5.2.1.2   Image Steganography

Pixels are the building blocks of an image. Every pixel contains three values (red, green, blue) also known as RGB values. Every RGB value ranges from 0 to 255. In this technique pixel intensities are used to hide the information. The 8 bit and 24 bit images are common. We can even hide an image inside another. An example of hiding data inside an image is shown in Fig. 5.2.

5.2.1.3   Network Steganography/Protocol Steganography

Network protocol such as TCP, UDP, IP, etc. is used as carrier for steganography [7]. Both cryptography and steganography techniques are practically applied in imperfect communication environments imposed by physical features of information carriers. While this imperfectness is generally an obstacle for cryptography, it is an essential enabling condition for many network steganography techniques that utilize redundant communication protocols to cope with such imperfect environments to provide reliable communication. The storage method can hide the secret information both in the user data (payload) and in the protocol field (non-payload).

5.2.1.4   Audio Steganography

Taking audio as carrier for information hiding is called audio steganography. It is used for digital audio formats such as WAVE, MIDI, and AVI MPEG for steganography. The advantage of taking an audio file as a carrier is that there is a large data redundancy, which means it can be embedded in a large amount of information [8].

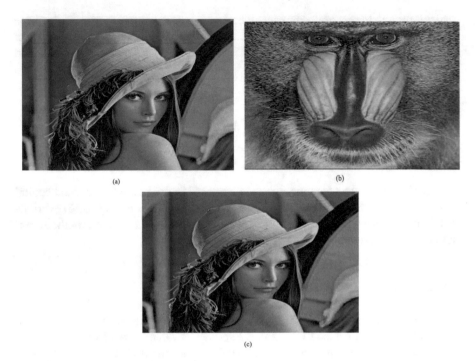

Fig. 5.2 (**a**) Raw image, (**b**) image to be hided, (**c**) stegano-image

Different methods of audio steganography are used such as least significant bit encoding, parity encoding, phase coding, and spread spectrum.

5.2.1.5   Video Steganography

It is a technique to hide any type of files or information into digital video format (H.264, Mp4, MPEG, and AVI). The combination of pictures is used as carrier for hidden information [9]. Video steganography is becoming an important research area in various data hiding technologies, which has become a promising tool because not only the security requirement of secret message transmission is becoming stricter but also video is more favored. Two main classes of video steganography are embedding data in uncompressed raw video and compressing it later and embedding data directly into the compressed data stream.

5.2.2   Steganography Techniques

Image steganography is broadly categorized into two categories, viz., spatial domain image steganography and frequency domain image steganography. While in former the message image is simply hidden in the spatial domain of the base image, i.e., no changes are to be done on the cover image for this domain, the latter cover image is converted into frequency domain by using discrete cosine transform, discrete wavelet transform, or any other frequency transformation mechanism, and then the message image is hidden into the coefficients, and further inverse is carried out which is to be transmitted.

5.2.2.1 Least Significant Bit (LSB)

LSB is one the technique of spatial domain methods. Changes in the value of the LSB are imperceptible for human eyes. The image obtained after embedding is almost similar to original image because the change in the LSB of image pixel does not bring too much differences in the image. A digital image consists of a matrix of color and intensity values. In a typical gray-scale image, 8 bits/pixel are used. In a typical full-color image, there are 24 bits/pixel, 8 bits assigned to each color components. Using 8 bit image, the least significant bit (the 8th bit) of some or all of the bytes inside an image is changed to a bit of the secret message. Using 24 bit image, a bit of each of the red, green, and blue color components can be used [10].

5.2.2.2 The Binary Pattern Complexity (BPC)

BPC approach is used to measure the noise factor in the image complexity. The noisy portion is replaced by binary pattern, and it is mapped from the secret data.

5.2.2.3 Discrete Wavelet Transform (DWT)

It is used to transform the image from a spatial domain to the frequency domain. In the process of steganography, DWT identifies the high-frequency and low-frequency information of each pixel of the image. It is a mathematical tool for decomposing an image hierarchically. It is mainly used for processing of non-stationary signals. The wavelet transform is based on small waves, known as wavelets, of different frequency and limited duration.

5.2.2.4 Spread Spectrum (SS)

The concept of spread spectrum is used in this technique. In this method the secret data is spread over a wide frequency bandwidth. The ratio of signal to noise in every frequency band must be so small that it becomes difficult to detect the presence of data. Even if parts of data are removed from several bands, there would be still enough information present in other bands to recover the data. Thus it is difficult to remove the data completely without entirely destroying the cover. It is a very robust approach used in military [11].

5.2.2.5   Multilevel Steganography (MLS)

In MLS, at least two steganography methods are utilized simultaneously, in such a way that one method (called the upper-level) serves as a carrier for the second one (called the lower-level). This makes the steganogram unreadable [12].

5.2.2.6   Hybrid Cryptography and Steganography

The data will be encrypted using RSA algorithm or any encryption technique. On the other hand, the encrypted data will be hided in image files or any other file formats.

5.2.3   Steganography Metrics

- *Robustness:* the ability of embedded data to remain intact if the stego-image undergoes transformations, such as linear and nonlinear filtering, sharpening or blurring, addition of random noise, rotations and scaling, cropping or decimation, and lossy compression.
- *Imperceptibility:* The imperceptibility means invisibility of a steganography algorithm. Because it is the first and foremost requirement, the strength of steganography lies in its ability to be unnoticed by the human eye.
- *Bit Error Rate:* The hidden information can be successfully recovered from the communication channel.
- *Mean Square Error:* It is computed by performing byte-by-byte comparisons of the two images.
- *Peak Signal-to-Noise Ratio (PSNR):* The image steganography system must embed the content of hidden information in the image so that the quality of the image should not change.

5.3   Watermarking

Watermarking is embedding some information within a digital media so that the digital media looks unchanged. The hidden information itself is not important by itself, but it says something about the digital media [13, 14]. It is a concept closely related to steganography, in that they both hide a message inside a digital signal. However, what separates them is their goal. Watermarking is used for *authenticity* (proof of ownership) [15]. The process of watermarking is divided into two parts: embedding of watermark into host image and extraction of watermark from image [16]. Watermarking uses the same techniques used by steganography [17]. An example of watermarking is shown in Fig. 5.3. A comparison between cryptography and watermarking is shown in Table 5.2.

Fig. 5.3 (**a**) Original image, (**b**) watermarked image

Table 5.2 Comparison between watermarking and cryptography [18–20]

Comparison	Cryptography	Watermarking
Basic	It means secret writing	It is known as cover writing
Goal	Data protection	Copyright protection
Structure of the message	Altered only of the transmission	Not altered
Popularity	More commonly used	Less popular
Relies on	No parameters	Key
Supported security principles	Confidentiality, data integrity, authentication, and non-repudiation	Confidentiality and authentication
Techniques	Transposition, substitution, stream cipher, block ciphers	Application-based: Using a universal logo without encryption in the embedding algorithm
Implemented on	Only on text files	Audio, video, image, text
Types of attack	Steganalysis	Cryptanalysis

5.3.1 Watermarking Metrics

- *Effectiveness:* This is the most important property of watermark that the watermark should be effective means it should surely be detective. If this will not happen, the goal of the watermarking is not fulfilled [21].
- *Host signal quality:* This is also an important property of watermarking. Everybody knows that in watermarking, watermark is embedded in host signal (image, video, audio, etc.). This may put an effect on the host signal. So the watermarking system should be like as, it will minimum changes the host signal and it should be unnoticeable when watermark is invisible [22].
- *Watermark size:* Watermark is often used for owner identification or security confirmation of host signal, and it is always used when data is transmitted. So it

is important that the size of watermark should be minimum because it will increase the size of data to be transmitted.

- *Robustness:* Robustness is a crucial property for all watermarking systems. There are so many causes by which watermark is degraded, altered during transmission, and attacked by hackers in paid media applications. So watermark should be robust, so that it withstands against all the attacks and threats.

5.3.2   Watermarking Applications

- *Owner Identification:* The application of watermarking to which he developed is to identify the owner of any media. Some paper watermark is easily removed by some small exercise of attackers. So the digital watermark was introduced. In that the watermark is an internal part of digital media so that it cannot be easily detected or removed. Embedding a strong watermark is important to prove IP core ownership during conflict resolution process [23].
- *Copy Protection:* Illegal copying is also prevented by watermarking with copy protect bit.

For example, professional photographers tend to watermark digital proofs sent to clients until the client agrees to purchase the photos, at which the original, unaltered images are released. This allows the photographer to distribute demos and samples of their work, without actually giving the original compositions. Moreover, it can be used for IP protections such as HDL IPs [24].

- *Broadcast Monitoring:* Broadcasting of TV channels and radio news is also monitoring by watermarking. It is generally done with the paid media like sports broadcast or news broadcast.
- *Data Authentication:* Sender embedded the digital watermark with the host data, and it would be extracted at the receivers end and verified.

5.4   Visual Cryptography

Like steganography, visual cryptography is used for hiding secret information but unlike steganography which can hide data in files, images, audio, and video. Visual cryptography hides it in multiple images. Visual cryptography is more secured against various attacks than steganography, and it does not require complex computation. In visual cryptography, each pixel in the original secret image can be expanded to any number of subpixels. After expansion, each pixel would be having a total number of "m" subpixels [25–27]. Visual cryptography is a cryptographic technique which allows visual information (pictures, text, etc.) to be encrypted in such a way that the decrypted information appears as a visual image. Visual cryptography can be used to protect biometric templates in which decryption does not require any complex computations.

5.5 Conclusions

In this chapter, different steganography and watermarking techniques and methods are introduced and analyzed. Steganography is the art and science of covered or hidden writing. The purpose of steganography is covert communication to hide the existence of a message from the eyes. Multilevel steganography is used to improve secret communication. Moreover, this chapter gives a deep introduction of watermarking. Both steganography and watermarking use steganography techniques to embed data covertly in noisy signals. While steganography aims for imperceptibility to human eyes, watermarking tries to control the robustness.

References

1. https://asmp-eurasipjournals.springeropen.com/articles/10.1186/1687-4722-2012-25/figures/1
2. A.K. Bairagi, R. Khondoker, R. Islam, An efficient steganographic approach for protecting communication in the Internet of Things (IoT) critical infrastructures. Inf. Sec. J. Glob. Perspect. **25**(4–6), 197–212 (2016)
3. W. Tang, B. Li, S. Tan, M. Barni, J. Huang, CNN-based adversarial embedding for image steganography. IEEE Trans. Inform. Foren. Sec. **14**(8), 2074–2087 (2019)
4. Almaadeed, N., Elharrouss, O., Al-Maadeed, S., Bouridane, A. and Beghdadi, A., 2019. A Novel Approach for Robust Multi Human Action Detection and Recognition Based on 3-Dimentional Convolutional Neural Networks. arXiv preprintarXiv:1907.11272
5. https://onlinelibrary.wiley.com/doi/full/10.1002/sec.1752
6. K. Rabah, Steganography-the art of hiding data. Inf. Technol. J. **3**(3), 245–269 (2004)
7. O.I. Abdullaziz, V.T. Goh, H.-C. Ling, K. Wong, Network packet payload parity based steganography, in *Sustainable Utilization and Development in Engineering and Technology (CSUDET) 2013 IEEE Conference*, (2013), pp. 56–59
8. https://aip.scitation.org/doi/pdf/10.1063/1.5039018
9. https://www.sciencedirect.com/science/article/abs/pii/S0925231218312608
10. S. Changder, N.C. Debnath, D. Ghosh, A new approach to Hindi text steganography by shifting matra, in *Advances in Recent Technologies in Communication and Computing, 2009. ARTCom'09. International Conference*, (2009), pp. 199–202
11. https://www.researchgate.net/publication/5583605_Spread_spectrum_image_steganography
12. W. Frączek, W. Mazurczyk, K. Szczypiorski, Multi-level steganography applied to networks, in *Proc. of: Third International Workshop on Network Steganography (IWNS 2011) co-located with The 2011 International Conference on Telecommunication Systems, Modeling and Analysis (ICTSM2011), Prague, Czech Republic*, (2011)., 27–28 May 2011
13. K. Rajitha, U.R. Nelakuditi, V.N. Mandhala, T.-H. Kim, FPGA implementation of watermarking scheme using XSG. Int. J. Sec. Its Appl. **9**(1), 89–96 (2015)
14. R.M. Khoshki, Hardware based implementation of an image watermarking system. Int. J. Adv. Res. Comp. Commun. Eng. **3**(5) (2014)
15. J. Fridrich, M. Goljan, R. Du, Detecting LSB steganography in color, and gray-scale images. IEEE Multimedia, 22–28 (n.d.)
16. P. Singh, R.S. Chadha, A survey of digital watermarking techniques, applications and attacks. Int. J. Eng. Innov. Tech. **2**, 9 (2013)
17. I.J. Cox, M.L. Miller, J.A. Bloom, J. Fridrich, T. Kalker, *Digital Watermarking and Steganography* (Morgan Kaufmann, 2008)
18. R. Bhandari, V.B. Kirubanand, Enhanced encryption technique for secure iot data transmission. **9**(5), 3732–3738 (2019)

19. S. Pramanik, R.P. Singh, R. Ghosh, A new encrypted method in image steganography. Indones. J. Electr. Eng. Comput. Sci. **14**(3), 1412 (2019)
20. R. Din, M. Mahmuddin, A.J. Qasim, Review on steganography methods in multi-media domain. Int. J. Eng. Technol. **8**(1.7), 288–292 (2019)
21. Z. Jalil, A.M. Mirza, A review of digital watermarking techniques for text documents, in *Proceedings of the Proceeding of the International Conference on Information and Multimedia Technology (ICIMT'09)*, (2009), pp. 230–234
22. M. Pal, A survey on digital watermarking and its application. Int. J. Adv. Comput. Sci. Appl. **7**(1), 153–156 (2016)
23. M.A. Qadir, I. Ahmad, Digital text watermarking: Secure content delivery and data hiding in digital documents. IEEE Aerosp. Electron. Syst. Mag. **21**(11), 18–21 (2006)
24. https://link.springer.com/chapter/10.1007/978-3-540-30114-1_16
25. M. Naor, A. Shamir, Visual cryptography. Adv. Cryptogr., 1–12 (1995)
26. Shiny, R. M., et al. "An efficient tagged visual cryptography for color images." 2016 IEEE International Conference on Computational Intelligence and Computing Research (ICCIC). IEEE
27. K. Dhiman, S.S. Kasana, Extended visual cryptography techniques for true color images. Comp. Elect. Eng. **70**, 647–658 (2018)

Chapter 6
Conclusions

In this book, the fundamentals of cryptography are discussed. It provides a comprehensive study of the three critical aspects of security: confidentiality, integrity, and authentication. As it is known, cryptography plays a vital and critical role in achieving the primary aims of security goals, such as authentication, integrity, confidentiality, and no-repudiation. Cryptographic algorithms are developed in order to achieve these goals as discussed during this book. Moreover, this book discusses the fundamentals of private and public key cryptography. Moreover, it explains the details of the main building blocks of these cryptographic systems. Besides, this chapter explores the different crypto-analysis techniques. It addresses stream ciphers, DES and 3DES, AES, block ciphers, the RSA cryptosystem, and public key cryptosystems based on the discrete logarithm problem, ECC, key exchange algorithms, and so many other algorithms. Moreover, this chapter provides a comparison between different encryption algorithms in terms of speed encryption, decoding, complexity, the length of the key, structure, and flexibility. This book explores different cryptography concepts such as authentication, integrity, availability, access control, and non-repudiation. It presents concepts of digital signatures, hash functions, and MACs. Finally, this book discusses the cutting-edge cryptographic techniques such as quantum cryptography, DNA cryptography, chaotic cryptography, lightweight cryptography, and blockchain cryptography. All these cryptography techniques are promising and rapid emerging fields in data security.

© The Editor(s) (if applicable) and The Author(s), under exclusive license to
Springer Nature Switzerland AG 2020
K. S. Mohamed, *New Frontiers in Cryptography*,
https://doi.org/10.1007/978-3-030-58996-7_6

Index

© The Editor(s) (if applicable) and The Author(s), under exclusive license to
Springer Nature Switzerland AG 2020
K. S. Mohamed, *New Frontiers in Cryptography*,
https://doi.org/10.1007/978-3-030-58996-7

Printed in the United States
by Baker & Taylor Publisher Services